Intracellular pH and Its Measurement

Authors

Arnošt Kotyk, Ph.D., D.Sc.
and
Jan Slavík, Ph.D.

Institute of Physiology
Czechoslovak Academy of Sciences
Prague, Czechoslovakia

CRC Press, Inc.
Boca Raton, Florida

Library of Congress Cataloging-in-Publication Data

Kotyk, Arnošt.
Intracellular pH and its measurement.

Bibliography: p.
Includes index.
1. Cytochemistry--Methodology. 2. Hydrogen-ion concentration--Measurement. I. Slavík, Jan
II. Title.
QH611.K67 1989 574.87′6042 88-19253
ISBN 0-8493-4916-8

This book represents information obtained from authentic and highly regarded sources. Reprinted material is quoted with permission, and sources are indicated. A wide variety of references are listed. Every reasonable effort has been made to give reliable data and information, but the author and the publisher cannot assume responsibility for the validity of all materials or for the consequences of their use.

Direct all inquiries to CRC Press, Inc., 2000 Corporate Blvd., N.W., Boca Raton, Florida, 33431.

International Standard Book Number 0-8493-4916-8

Library of Congress Card Number 88-19253
Printed in the United States

PREFACE

At the very beginning of my scientific career more than 30 years ago, I was interested in shifts of various phosphate fractions of yeast under different physiological conditions; the measurements included the effects of various ions on these fractions. In several experiments where a Na^+-containing medium was compared with a K^+-containing one there was a strikingly higher level of inorganic orthophosphate in the yeast with K^+ than there was with Na^+. We invented several working hypotheses and set out to test them. When I checked the bottles of Na- and K-phosphates we had used, sure enough, one was Na_2HPO_4 and the other KH_2PO_4, with the pH of the somewhat acid-buffered medium being 8.8 and 6.1, respectively. The observed effect was solely due to the difference in pH.

This was my first practical lesson on the importance of pH in biology. Others followed later after it had become clear that the intracellular (as well as the extracellular) pH is a quantity that affects the kinetics and the energetics of many transport systems and, finally, when the local pH — whatever its interpretation may be — was found to differ quite generally from the bulk pH.

The realization that H^+ ions not only play a role in enzyme (and transport) catalysis but also are involved in the energization of some transport processes, has led to a burst of emphasis on accurate measurements of their concentrations, a task seemingly quite straightforward but, in practice, hampered by both conceptual and technical problems.

We attempted to summarize in this book the present state-of-the-art of pH measurement, with a special accent on intracellular pH, proceeding from the rather extensive experience with several methods of the senior author (A.K.), and the intensive experience with the dual fluorescence method invented by the junior author (J.S.), the method being the one of choice in general laboratory practice where local pH values are sought.

We had the opportunity to draw on the know-how of many colleagues, some of whom have read and corrected parts of the book; thus, we wish to acknowledge the following people: Prof. J. Koryta of the J. Heyrovský Institute of Physical Chemistry and Electrochemistry in Prague (Chapters 1 and 2), Dr. J. Buděšínský of the Institute of Organic Chemistry and Biochemistry in Prague, Dr. P. Sedmera of the Institute of Microbiology in Prague, Dr. I. Goljer of the Slovak Technical College in Bratislava (Chapter 4E), and Dr. J. Plášek of the Charles University of Prague (Chapters 4C and 4D). We also wish to thank the 75 authors, quoted at the appropriate places of the book, who sent their reprints for our perusal.

Arnošt Kotyk
Jan Slavík

THE AUTHORS

Arnošt Kotyk, Ph.D., D.Sc., is a senior scientist at the Institute of Physiology, Czechoslovak Academy of Sciences in Prague, Czechoslovakia.

Dr. Kotyk received his various academic degrees at the University of California in Berkeley, at the Charles University in Prague, and at the Institute for Organic Chemistry and Biochemistry also in Prague, between the years 1950 and 1958.

Dr. Kotyk is the chairman of the State Research Board on Special Cell Biology, lectures at the J. E. Purkyně University in Brno and at the Charles University in Prague, and holds important positions in the International Union of Biochemistry and in the UNESCO Global Program on Molecular and Cellular Biology.

Dr. Kotyk has published 240 original papers, has written six monographs from the field of membrane transport and enzyme kinetics, and has lectured in 26 countries in five languages.

Jan Slavík, Ph.D., is a Research Scientist at the Institute of Physiology, Czechoslovak Academy of Sciences in Prague. He majored in biophysics and received his degrees at the Charles University in Prague.

Dr. Slavík is the leader of a group researching fluorescence techniques at the Institute of Physiology of the Czechoslovak Academy of Sciences and has published over 25 original papers and several reviews for international journals.

Dr. Slavík's current interests include localization and interactions of membrane transport proteins.

TABLE OF CONTENTS

Chapter 1

THE CONCEPT OF pH, HYDROLYSIS, AND BUFFERS

Arnošt Kotyk

TABLE OF CONTENTS

I. DEFINITION OF pH

Some 80 years ago, Sörensen introduced the symbol pH for the negative logarithm of concentration (moles or equivalents per cubic decimeter) of H^+ ions.[1] The letter "p" derives either from the Latin *pondus* (weight or significance) or from *potentia* (might or potential) and has since been used to denote the negative logarithm of concentration of various other ions; e.g., $\log c_{Na+} = -pNa$, which is correct or at least consistent. It has also been used for the negative logarithms of various kinetic quantities having dimensions of concentration, such as the dissociation constant K_a, where $-\log K_a = pK_a$, which is certainly incorrect in the original sense of "p" but is useful in cases where sums or differences, such as pK_a − pH, occur in formulae.

In Sörensen's original version

$$pH = -\log_{10} c_{H^+} \tag{1}$$

but it is now defined in terms of activities a. Thus (with γ being the activity coefficient),

$$pH = -\log_{10} (c_{H^+} \gamma_{H^+}) = -\log a_{H^+} \tag{2}$$

which, particularly in more concentrated solutions, represents an appreciable difference, as will be seen below.

It should be observed that in rigorous derivations of these equations both c and a (and, hence, γ) are dimensionless quantities that can be understood as concentrations or activities relative to a unit reference. Their values are thus numerically identical with true concentrations or activities expressed in moles per cubic decimeter.

Water, which is the principal milieu of all subcellular and cellular systems, is an extremely weak electrolyte; but it is still dissociated into ions, according to

$$H_2O \rightleftharpoons H^+ + OH^- \tag{3}$$

As the H^+ ion cannot exist as free proton in aqueous media, the equation should be properly written with an oxonium ion in place of H^+. Thus,

$$2H_2O \rightleftharpoons H_3O^+ + OH^- \tag{4}$$

Even in an acid solution of pH = 0 the concentration of free protons was shown to be about 10^{-130}![2] However, this way of expressing the hydrogen ion concentration does not reflect its actual solvation degree in water which, in fact, shows highest stability for $H_3O^+{\cdot}3H_2O$ and for $OH^-{\cdot}3H_2O$.

From Equation 4, then, the dissociation constant of water is

$$K_A = a_{H_3O^+} \cdot a_{OH^-}/a_{H_2O} = c_{H_3O^+}\gamma_{H_3O^+} \cdot c_{OH^-}\gamma_{OH^-}/c_{H_2O}\gamma_{H_2O} \tag{5}$$

The dissociation of water being negligible we may consider the concentration of H_2O as constant and the activity coefficients equal to unity. The concentration of water itself is 55.35 mol dm^{-3} at 25°C and it is generally included in what is termed the ionization constant (ion product) of water, or

$$K_w = a_{H_3O^+} \cdot a_{OH^-} \tag{6}$$

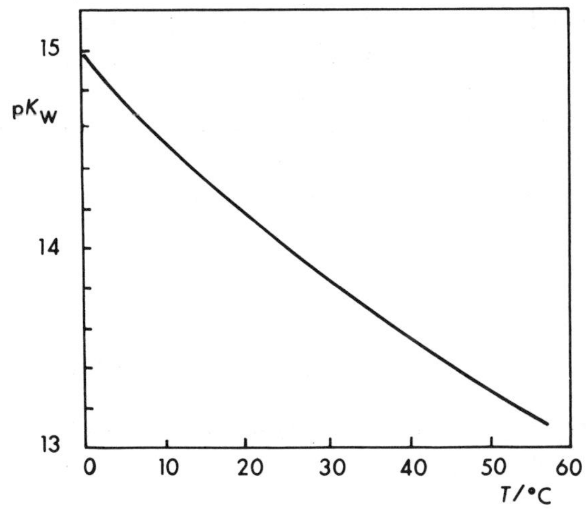

FIGURE 1. Temperature dependence of the ionization constant (ion product) of water.

The value of K_w is almost exactly 10^{-14} mol dm^{-3} at 25°C and depends on temperature as shown in Figure 1. The degree of dissociation of water α ($= c_{H_3O^+}/c_{H_2O} = c_{OH^-}/c_{H_2O}$) at 25°C is, thus, $1.41 \cdot 10^{-9}$.

Because of the temperature dependence, neutrality (identical number of plus and minus charges in solution) is achieved at 6.81 at human body temperature.

For the sake of comparison, it should be noted that the self-ionization (autoprotolysis) constants of protogenic solvents (strong acids) are high ($2.69 \cdot 10^{-4}M$ for sulfuric acid), but much like water for weaker acids or alcohols ($3.16 \cdot 10^{-15}M$ for acetic acid and $2.51 \cdot 10^{-17}M$ for ethanol), and are extremely low for protophilic solvents (bases), e.g., $10^{-29}M$ for ammonia.

II. IONIZATION OF ACIDS AND BASES

Although water is by far the most abundant component of all living systems, its dissociation into oxonium ions is so weak that the pH of a salt solution, either extra- or intracellular, is determined by the presence of components that readily dissociate or readily bind an oxonium ion, i.e., acids and bases, respectively.

Throughout modern electrochemistry, three theories of acids and bases came into prominence.

The first theory, that of Arrhenius[3] dates back to 1887 when he postulated the universal existence of dissociation of electrolytes in solution, supporting his views by conductometric measurements. He calculated the degree of dissociation α from the ratios of equivalent conductivities at a given and at an infinite dilution. Thus,

$$\alpha = \Lambda/\Lambda_\infty$$

It was Arrhenius who defined acids and bases in a simple way, stating that an acid (HA) is characterized by dissociation of hydrogen ions

$$HA \rightleftharpoons H^+ + A^- \tag{7a}$$

while a base (BOH) is recognized as a substance dissociating hydroxide ions

$$BOH \rightleftharpoons B^+ + OH^- \tag{7b}$$

Although many of his deductions are still valid, particularly with respect to aqueous solutions, he could not foresee the behavior of acids and bases in nonaqueous solvents and the role of interactions between the solute and the solvent.

The second theory is the one that is most relevant to our considerations of pH in aqueous solutions and is due to Brönsted[4] who replaced Arrhenius' definition of acids and bases by stating more generally that an acid is any substance that can dissociate a proton; a base is then any substance that can bind a proton. Thus,

$$A \rightarrow B + H^+, \quad \text{e.g.,} \quad HNO_3 \rightarrow NO_3^- + H^+ \tag{8a}$$

$$B + H^+ \rightarrow A, \quad \text{e.g.,} \quad C_6H_5NH_2 + H^+ \rightarrow C_6H_5NH_3^+ \tag{8b}$$

Clearly, such reactions (much like analogous oxidation-reductions where an electron is exchanged) can only proceed jointly, such as

$$AH + B \rightarrow A + BH \tag{9}$$

This pattern comprises a variety of H^+ exchanges, such as the above-mentioned self-dissociation of water

$$H_2O + H_2O \rightarrow OH^- + H_3O^+ \tag{10a}$$

or salt formation

$$HBr + NH_3 \rightarrow Br^- + NH_4^+ \tag{10b}$$

or dissociation of an acid

$$H_3PO_4 + H_2O \rightarrow H_2PO_4^- + H_3O^+ \tag{10c}$$

or of a base

$$H_2O + NH_3 \rightarrow OH^- + NH_4^+ \tag{10d}$$

etc.

Brönsted's principal contributions to the theory were numerous.

1. He proposed that both ions and undissociated molecules can act as proton donors or acceptors.
2. He introduced the idea that suitable substances can act as acids or bases even in nonaqueous media.
3. He suggested that the dissociation process of an acid and the association process of a base are both governed by the nature of the solvent so that a given substance can act either as a base or as an acid, depending on the type of the solvent. Thus, if the affinity of the dissolved substance (solute) for the proton is greater than the affinity of the solvent anion for the proton, the solute behaves as a base (e.g., aniline in water but also acetic acid in concentrated sulfuric acid). On the other hand, if the proton affinity of the "base" conjugated with the solute is less than the affinity of solvent molecules for the proton, the solute acts as an acid (e.g., acetic acid in water but also aniline in liquid ammonia).

4. He pointed out that the amphoteric character of the solvent underlies its ability to undergo self-ionization.
5. He demonstrated that hydrogen ions in solution are solvated by the molecules of the solvent, thus, H_3O^+ and not H^+, or NH_4^+ and not H^+.

Modern classifications of solvents are, in fact, based on Brönsted's concepts.

Protic solvents are able to dissociate protons, either in a clear-cut fashion (in which case solvents are *protogenic,* such as strong acids) or in an amphoteric fashion (where solvents are *amphiprotic,* such as water itself).

Aprotic solvents cannot dissociate protons. They are either *protophilic,* i.e., they can bind protons (e.g., dimethyl sulfoxide), or they are completely inert toward protons (e.g., benzene).

The third theory describing the behavior of acids and bases was proposed by G. N. Lewis and is even more general than that of Brönsted in defining a base as any electron-pair donor and an acid as any electron-pair acceptor. According to this concept acids and oxidizing agents can be placed together in the group of electrophilic substances while bases and reducing agents are nucleophilic. While this all-embracing approach is most rigorous as far as the actual electron configurations of the reacting substances are concerned, for our present purposes, i.e., definition and implications of pH, it is of little significance. Thus, according to this view, the so-called *hard acids* include not only H^+, but also Sn^{2+}, I^{5+}, CO_2, and SO_3, and the *soft acids* include no ion or molecule that would be able to dissociate a proton. His *hard bases* are not only OH^-, but also H_2O, ethanol, and diethyl ether, with *soft bases* comprising not only I^- and CN^-, but also CO, ethylene, benzene, etc.[5]

For the present purposes, the framework of Brönsted's theory is superior to any other, and we shall proceed to show on the basis of this theory how the pH of a solution is related to the quality and the quantity of acids and bases dissolved in it.[6]

It should be noted at this point that in the following two kinds of electrolytes will be dealt with: strong and weak. Strong electrolytes are those that are fully dissociated in solution and may comprise strong acids (such as H_2SO_4) and strong bases (such as NaOH) as well as salts composed of at least one strong partner. Thus, not only Na_2SO_4, but also sodium acetate and manganous sulfate will be fully dissociated in an aqueous solution. (This should not be confused with the restricted solubility of some such compounds, e.g., AgCl.)

Weak electrolytes are, above all, weak acids (such as carbonic acid), weak bases (such as hydrazine), and, obviously, salts of such weak partners.

The fundamental formula and equations will all operate with concentrations (mol dm^{-3}) rather than activities, because these can always replace concentrations if one multiplies the appropriate values with activity coefficients, as long as these are amenable to measurement. That is why, for instance, the dissociation constants shown here are actually apparent dissociation constants that may differ substantially from the true, thermodynamic dissociation constants based on activities. These apparent dissociation constants are often distinguished by a prime from the true ones (K' vs. K), but the simple notation will be used throughout.

A *weak monobasic acid,* such as are mostly encountered in biological systems, dissociates in water according to

$$HA + H_2O \rightleftharpoons A^- + H_3O^+ \tag{11}$$

with the corresponding dissociation constant being defined as

$$K_A = c_{H_3O^+} \cdot c_{A^-}/c_{H_2O} = c_T\alpha^2/(1 - \alpha) \tag{12}$$

where $c_T = c_{A^-} + c_{HA}$ and α, the degree of ionization, is equal to c_{A^-}/c_T or $c_{H_3O^+}/c_T$. This

This degree increases toward unity as c_T approaches zero. (With strong acids $\alpha \cong 1$ even at moderate concentrations.)

From the above equation

$$c_{H_3O^+} = c_F\alpha = {}^1\!/_2 \left(-K_A + \sqrt{K_A^2 + 4c_TK_A}\right) \tag{13}$$

For weak acids, at medium concentrations $\alpha \ll 1$ so that $c_T = c_{HA}$ and

$$c_{H_3O^+} \cong \sqrt{c_TK_A} \tag{14a}$$

or

$$\text{pH} \cong {}^1\!/_2 \left(\text{p}K_A - \log c_A\right) \tag{14b}$$

The use of this relatively simple formula is justified (i.e., the error of determination is less than 1%) only with certain limitations. The value of $c_{H_3O^+}$ defined by Equation 14a differs from that defined by Equation 13 in the expression ${}^1\!/_2\sqrt{K_A/c_{T'}}$ as may be shown by the following consideration. It follows from Equation 12 that

$$\alpha^2 = K_A(1 - \alpha)/c_T \tag{15a}$$

In the first approximation (for α very small), then,

$$\alpha^2 = K_A/c_T \tag{15b}$$

while in the second approximation

$$\alpha^2 = (K_A/c_T)\left(1 - \sqrt{K_A/c_T}\right) \tag{15c}$$

For $\sqrt{K_A/c_T}$ to be less than 0.01 in the expression for α, it must be approximately less than 0.02 in the expression for α^2, so that K_A/c_T must be less than $4 \cdot 10^{-4}$. Hence, for an acid with a dissociation constant of $8.7 \cdot 10^{-4}$ mol dm^{-3}, such as citric acid, a more than 2 *M* concentration would be required to justify the use of the simplified Equation 14a. It is only with weak acids, such as carbonic or acetic, or with the further dissociation constants of di- and tri-basic acids that the simplification is in order.

A *weak dibasic* acid will dissociate in two steps:

$$H_2A + H_2O \rightleftharpoons HA^- + H_3O^+ \tag{16a}$$

$$HA^- + H_2O \rightleftharpoons A^{-2} + H_3O^+ \tag{16b}$$

with the corresponding dissociation constants

$$K_{A_1} = c_{HA^-} \cdot c_{H_3O^+}/c_{H_2A} = c_T\alpha_1^2(1 - \alpha_2^2)/(1 - \alpha_1) \tag{17a}$$

$$K_{A_2} = c_{A^{2-}} \cdot c_{H_3O^+}/c_{HA^-} = c_T\alpha_1\alpha_2(1 + \alpha_2)/(1 - \alpha_2) \tag{17b}$$

where $c_T = c_{H_2A} + c_{HA^-} + c_{A^{2-}}$.

If again $\alpha_{1,2} \ll 1$, the last two equations simplify to

$$K_{A_1} \cong c_1\alpha_1^2 \tag{18a}$$

$$K_{A_2} \cong c_1\alpha_1\alpha_2 \tag{18b}$$

Since $c_{H_2A} = c_T(1 - \alpha_1)$, $c_{HA^-} = c_T\alpha_1(1 - \alpha_2)$, and $c_{A^{2-}} = c_T\alpha_1\alpha_2$, it follows that

$$c_{H_3O^+} = c_T\alpha_1(1 + \alpha_2) \cong \sqrt{K_{A_1}c_T} + K_{A_2} \tag{19}$$

If $K_{A_1} \gg K_{A_2}$, Equation 19 simplifies to Equation 14a.

If, however, K_{A_1} is similar to K_{A_2}, $c_{H_3O^+}$ must be calculated from the cubic equation

$$c_{H_3O^+}^3 + K_{A_1}c_{H_3O^+}^2 - (K_{A_1}c_T - K_{A_1}K_{A_2})c_{H_3O^+} = 2K_{A_1}K_{A_2}c_T \tag{20}$$

To determine the pH of a solution of a *weak base* we proceed from its dissociation

$$B + H_2O \rightleftharpoons BH^+ + OH^- \tag{21}$$

and the corresponding dissociation constant K_B

$$K_B = c_{BH^+} \cdot c_{OH^-}/c_B = \alpha^2 c_T/(1 - \alpha) \tag{22}$$

Using the ionization constant of water $K_W = c_{H_3O^+} \cdot c_{OH^-}$ and keeping in mind that $c_T = c_B + c_{BH^+}$, we obtain under analogous simplifying assumptions ($\alpha \ll 1$) that

$$c_{H_3O^+} \cong K_W/\sqrt{K_B c_T} \tag{23a}$$

or

$$pH \cong pK_W + {}^1/_2\,(\log c_T - pK_B) \tag{23b}$$

It is now common use that K_B is replaced with K_{BH} where

$$K_{BH} = c_B \cdot c_{H_3O^+}/c_{BH^+} \tag{24}$$

from

$$BH^+ + H_2O \rightleftharpoons B + H_3O^+ \tag{25}$$

The relationship between K_B and K_{BH} follows from

$$K_{BH} = K_W/K_B \tag{26a}$$

or

$$pK_{BH} = pK_W - pK_B \tag{26b}$$

Table 1
DISSOCIATION CONSTANTS OF SOME BIOLOGICALLY IMPORTANT ACIDS AND BASES IN WATER AT 25°C

Acid	pK_A	Base	pK_{BH+}
Malonic	2.85	Methylamine	10.66
Lactic	3.86	Ammonia	9.25
Butyric	4.82	Hydrazine	8.23
Sulfuric	Very	Quinine	
(1st)	low	(1st)	5.70
(2nd)	1.92	(2nd)	9.87
Oxalic			
(1st)	1.19	Aniline	4.63
(2nd)	4.11	α-Naphthylamine	3.92
Phosphoric			
(1st)	2.12		
(2nd)	7.21		
(3rd)	12.36		
Citric			
(1st)	3.06		
(2nd)	4.74		
(3rd)	5.40		
Ascorbic (1st)	3.24		
Succinic			
(1st)	4.16		
(2nd)	5.61		
Carbonic			
(1st)	6.37		
(2nd)	10.25		
Boric			
(1st)	9.14		
(2nd)	12.74		
(3rd)	13.80		

The dissociation constants of some weak acids and bases are shown in Table 1.

III. HYDROLYSIS OF SALTS

The pH of a salt solution depends on the relative strength of the acid and base components of the salt.

In the case of a *strong acid* and a *weak base,* such as NH_4NO_3, its dissociation leads to $NH_4^+ + NO_3^-$, the nitrate anion being an extremely weak base (conjugated with a strong acid HNO_3) and, hence, not participating in further reactions of protolysis. However, NH_4^+ acts as an acid in an aqueous solution as it is conjugated with a weak base (NH_3). It reacts with water according to

$$NH_4^+ + H_2O \rightleftharpoons H_3O^+ + NH_3 \qquad (27a)$$

or

$$BH^+ + H_2O \rightleftharpoons H_3O^+ + B \qquad (27b)$$

(One may neglect here the formation of very minor amounts of NH_4OH.) The resulting pH is necessarily acidic.

The *hydrolysis* constant is defined as

$$K_h = c_B \cdot c_{H_3O^+}/c_{BH^+} = K_W/K_B = \epsilon^2 c_T/(1 - \epsilon) \tag{28}$$

where ϵ is the degree of hydrolysis, such that $\epsilon = c_{H_3O^+}/c_T = c_B/c_T$; $(1 - \epsilon) = c_{BH^+}/c_T$, and $c_T = c_B + c_{OH^-}$. Then

$$c_{H_3O^+} = -K_{BH}[1 - \sqrt{K_W(1 + 4c_T/K_{BH})}] \tag{29}$$

For insignificant hydrolysis ($\epsilon \ll 1$)

$$c_{H_3O^+} \cong \sqrt{c_T K_{BH}} \tag{30a}$$

or

$$pH \cong {}^1/_2\,(pK_{BH} - \log c_T) \tag{30b}$$

If, on the other hand, hydrolysis is nearly complete so that all BH^+ added is converted into B and H_3O^+, then

$$pH \cong -\log c \tag{31}$$

In the case of a salt composed of a *weak acid* and a *strong base,* such as sodium acetate, the anion (acetate, A^-) is conjugated with a weak acid so that it acts as a base toward water, while the cation (Na^+, B) is not conjugated with a base and does not enter into protolytic reactions. The hydrolysis then proceeds according to

$$CH_3COO^- + H_2O \rightleftharpoons CH_3COOH + OH^- \tag{32a}$$

or

$$A^- + H_2O \rightleftharpoons HA + OH^- \tag{32b}$$

The resulting pH is necessarily alkaline. The hydrolysis constant is then

$$K_h = c_{HA} \cdot c_{OH^-}/c_{A^-} = K_W/K_A = \epsilon^2 c_T/(1 - \epsilon) \tag{33}$$

In analogy with the strong acid-weak base case one can derive (with $\epsilon = c_{OH^-}/c_T$ and $c_T = c_{HA^+}\ c_{A^-}$) that

$$c_{OH} = -K_W + \sqrt{K_W^2 + 4K_W K_A c_T}/K_A \tag{34a}$$

or

$$c_{H_3O^+} = -K_A(1 - \sqrt{4K_A c_T/K_W + 1}) \tag{34b}$$

which, for insignificant hydrolysis, turns into

$$c_{H_3O^+} \cong \sqrt{K_W K_A/c_T} \tag{35a}$$

or

$$pH \cong {}^1/_2\,(pK_w + pK_A + \log c_T) \tag{35b}$$

The most complicated case arises on dissolving a salt composed of a *weak acid* and a *weak base,* such as ammonium acetate. Both the dissociated ions, CH_3COO^- (= A^-) and NH_4^+ (= BH^+) undergo hydrolysis. Thus,

$$BH^+ + H_2O \rightleftharpoons B + H_3O^+ \tag{36}$$

with

$$K_{BH} = c_B \cdot c_{H_3O^+}/c_{BH^+} = K_W/K_B \tag{37}$$

and

$$A^- + H_2O \rightleftharpoons HA + OH^- \tag{38}$$

with

$$K_{A^-} = c_{HA} \cdot c_{OH}/c_{A^-} = K_W/K_A \tag{39}$$

In sum, then,

$$BH^+ + A^- \rightleftharpoons B + HA \tag{40}$$

the constant of hydrolysis being

$$K_h = c_B \cdot c_{HA}/c_{BH^+} \cdot c_{A^-} = K_{BH}/K_{A^-} = (\epsilon_A - \epsilon_{BH^+})/(1 - \epsilon_{A^-})(1 - \epsilon_{BH^+}) \tag{41}$$

In analogy with the above derivations, $c_B/c_T = \epsilon_{BH^+}$, $c_{HA}/c_T = \epsilon_{A^-}$, so that $c_{A^-}/c_T = 1 - \epsilon_{A^-}$ and $c_{BH^+}/c_T = 1 - \epsilon_{BH^+}$, with $c_T = c_{A^-} + c_{HA} = c_{BH^+} + c_B$. From this

$$c_{H_3O^+} = K_{BH}(1 - \epsilon_{BH^+})/\epsilon_{BH^+} = K_A\epsilon_{A^-}/(1 - \epsilon_{A^-}) \tag{42}$$

The full formula expressed in terms of dissociation constants and the total analytical concentration is rather unwieldy, but it simplifies for the case that hydrolysis is not extensive and of the same order of magnitude for both the acid and the basic components, whereupon

$$pH = {}^1/_2\,(pK_W + pK_A - pK_B) \tag{43}$$

In other words, the pH will be higher the weaker the acid and/or the stronger the base. It should be observed that under these simplifying conditions the pH of a solution of a salt composed of a weak acid and a weak base is independent of the salt concentration.

IV. BUFFERING POWER

For a mixture of dissolved salts and relatively weak acids, such as most biological fluids are, the pH is virtually impossible to calculate on the basis of their concentrations and dissociation and hydrolysis constants. Even if only the major ions are considered, their analytical concentrations may range over several orders of magnitude while the pH is maintained within a relatively narrow range. This is due to the buffering action of the mixture,

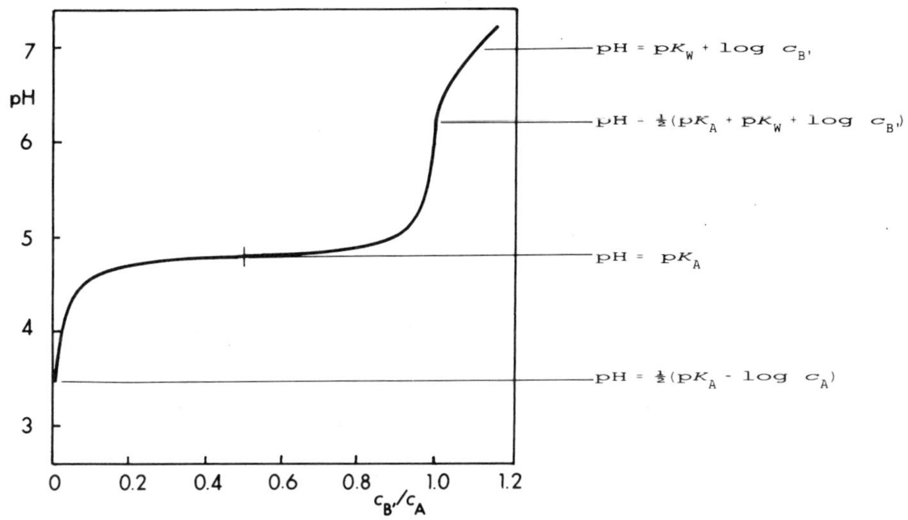

FIGURE 2. Titration of a weak acid A with a strong base B′. The pH observed at the different base-to-acid ratios is defined by the expressions on the right. The zone of high buffering power lies here at about 0.2 to 0.8.

a phenomenon that can be defined as the ability of a salt-and-acid- or salt-and-base mixture to withstand addition of strong acid or base so as to maintain its pH.

Let us consider a solution of a weak acid HA, such that $c_T = c_{HA} + c_{A^-}$. Its pH follows from the dissociation reaction (Equation 11) governed by K_A and can be expressed (for low degrees of dissociation) as

$$pH = pK_A + \log(c_{A^-}/c_{HA}) \tag{44}$$

(cf. Equation 12).

On adding a strong base to the solution, to a final concentration $c_{B'}$ a salt is formed at practically the same concentration so that $c_{B'} \cong c_{A^-}$. This necessarily leads to a decrease of the concentration of undissociated acid, such that $c_{HA} \cong c_T - c_{B'}$. This then turns into the so-called *Henderson-Hasselbalch* equation

$$pH = pK_A + \log[c_{B'}/(c_T - c_{B'})] \tag{45}$$

Analogously, starting with a weak base ($c_T = c_B + c_{BH^+}$) we may write

$$pH = pK_{BH} + \log(c_B/c_{BH^+}) \tag{46}$$

Addition of a strong acid to a final concentration $c_{A'}$ leads then to salt formation such that $c_{A'} \cong c_{BH^+}$ and, necessarily $c_B \cong c_T - c_{A'}$. The pH is then defined by

$$pH = pK_{BH} + \log[(c_T - c_{A'})/c_{A'}] \tag{47}$$

It is easy to show that the "titration" of, say, a weak acid with a strong base, proceeds through a "buffer zone" where pH changes very little in spite of adding large amounts of the base (Figure 2). The figure indicates that the buffer zone centers about a point where $pH = pK_{A'}$, this being also the point where $c_{B'} = {}^1/_2\, c_A$.

The efficiency of a buffer solution is usually expressed by its buffering power β which is the expression for a change of pH brought about by adding a certain amount of strong acid or base. Thus,

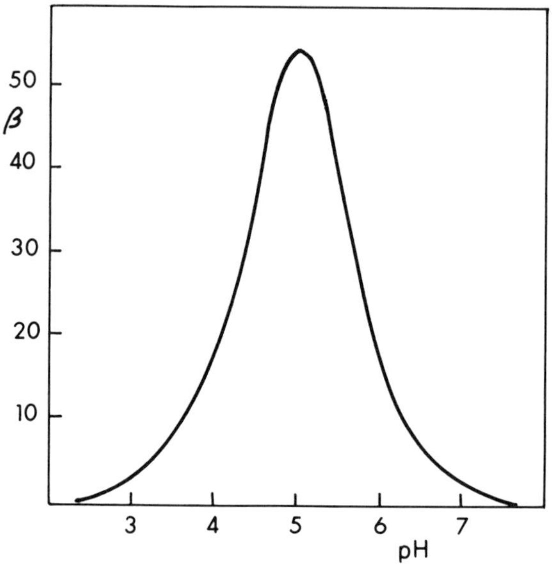

FIGURE 3. The buffering power β (in mmol dm^{-3}) of a weak acid buffer with pK_A = 5.

$$\beta = dc_{B'}/dpH = 2.3c_{B'}(1 - c_{B'}/c_A) \quad \text{for an acid buffer} \tag{48a}$$

$$\beta = dc_{A'}/dpH = 2.3c_{A'}(c_{A'}/c_B - 1) \quad \text{for an alkaline buffer} \tag{48b}$$

The buffering power thus depends on $c_{B'}$ or $c_{A'}$ and its maximum is reached (when the first derivative of Equation 48a or 48b is equal to zero) at $c_{B'} = {}^1/_2\, c_A$ or $c_{A'} = {}^1/_2\, c_B$ (Figure 3).

Buffer solutions are commonly used in biochemical and biological work, the endeavor being to use the same type of buffer (containing simply different proportions of acid and salt) for as wide a range of pH as possible. In preparing buffer solutions the ratio of salt:acid or salt:base usually ranges from 0.1 to 10, so that buffering is ensured over a range from pK_A − 1 to pK_A + 1 (or pK_{BH} + 1 to pK_{BH} − 1). However, at the limiting ratios the buffering power is only some 33% of the maximum.

A few words should be said about the buffering function of CO_2 which, particularly in biological systems, is ubiquitous and certainly forms a part of the overall buffering power of body fluids, including intracellular space. Carbon dioxide in the atmosphere, at concentrations ranging from 320 to 370 ppm, is in equilibrium with its dissolved counterpart in water through the absorption coefficient α_{CO_2} which, in milliliters CO_2 per milliliter H_2O, ranges from 0.878 at 20°C to 0.530 at 40°C, to be sure at a partial pressure of CO_2 of 100 kPa (now equal, by definition, to standard atmospheric pressure). At the actual partial pressure of CO_2 in normal atmosphere, equal to 35 Pa, the amount of dissolved CO_2 at 20°C would be 0.304 ml/l water. This CO_2 undergoes hydration

$$CO_2 + H_2O \rightleftharpoons H_2CO_3 \tag{49}$$

by a relatively slow process which can be speeded up in cells by the enzyme carbonate dehydratase. The equilibrium concentration of H_2CO_3 is given by

$$c_{H_2CO_3} = \alpha_{CO_2} p_{CO_2}/22.4p^{\oplus} \tag{50}$$

where $p_{CO_2}/p^{\oplus}$ is the fractional pressure of CO_2 and 22.4 is the volume of a mole of CO_2 in liters. From this, then, under normal laboratory atmosphere and at 20°C $c_{H_2CO_3}$ = $13.7 \mu M$.

If the dissolving and hydration of CO_2 proceeded ideally, a pure CO_2 atmosphere at normal pressure would bring $c_{H_2CO_3}$ to 39 mM. Carbonic acid is a very weak acid and is hydrolyzed readily to HCO_3^- and H_3O^+ with $pK_A = 6.37$. Since the second dissociation constant is practically negligible the first dissociation constant can be used to determine the pH of a H_2CO_3/CO_2 solution. From pH $\cong \frac{1}{2}(6.37 - \log c_{H_2CO_3})$ one will obtain a pH of 5.6 for laboratory-distilled water at 20°C (if no HCl is present) and a pH of 3.95 at 100 kPa CO_2 pressure and 30°C. Thus, "CO_2" buffers can be prepared directly from gaseous CO_2 and a suitable strong base, such as NaOH.

Buffers containing CO_2 and/or hydrogen carbonate are common in living systems and have often been imitated in preparing artificial bathing media for cells and tissues. Thus, in blood plasma the apparent concentration of H_2CO_3 is 1.25 mmol dm^{-3} and the corresponding concentration of (sodium) hydrogen carbonate must be (from Equation 45) 25 mmol dm^{-3} if the ambient pH is 7.4.

V. THE ACTIVITY COEFFICIENT

We have derived all the above equations using concentrations, not activities of the components. This is obviously not fully correct quantitatively although the equations will have the same form when activities are used consistently.

While in dilute solutions, in general, the activity coefficient (cf. Equation 2) is not very different from one, it may decrease in more concentrated solutions and particularly in a biological fluid. At moderate and high concentrations the deviation of the solution from ideality is more and more pronounced, and it appears that, due to interactions with the solvent and among themselves, the number of solvent particles that enter into reactions (i.e., are active) is distinctly less than would correspond to their analytical concentration. It is no mean task to derive the magnitude of activity coefficients accounting for this deviation from ideality, and it is certainly beyond the scope of this book — let alone the fact that three different activity coefficients are currently recognized in physical chemistry (γ_x — the rational coefficient corresponding to the mole fraction; γ_m — molality or practical coefficient; γ_c — molarity coefficient).

Let it only be said here that the derivation of the activity-coefficient value proceeds from determining the electrical work of charging the ion in solution when it is already surrounded by its ion atmosphere. This atmosphere is built up around the ion such that ions of the same sign are repelled from it and ions of the opposite sign are attracted to it. Thus, a region of space charge is formed around the ion which is exactly equal to the charge of the ion concerned. The activity coefficient corresponds to the work ΔG_{el} necessary for the transfer of the ion from the ideal situation where presumably no ion atmosphere exists around it into its position in the center of the ion atmosphere.

The classical equation based on the theory of Debye and Hückel[7] for a cation is

$$\log \gamma_+ = 0.43\Delta G_{el}/RT = -Az_+^2 \sqrt{I} \tag{51}$$

while for the mean activity coefficient of electrolyte $B_{\gamma+}A_{\gamma-}$, where $\gamma_\pm = (\gamma_+^{\nu+}\gamma_-^{\nu-})^{1/\nu}$,

$$\log \gamma_\pm = -Az_+z_- \sqrt{I} \tag{52}$$

where z is the electrical charge of the electrolyte, I is the ionic strength ($= \frac{1}{2}\sum_j c_jz_j$), and $A = F^3\sqrt{2\rho}/8\pi N_A \ln 10 \cdot \sqrt{\epsilon^3R^3T^3}$ with F (the Faraday constant) = 96.48 kC mol^{-1}, N_A =

Table 2
MEAN ACTIVITY COEFFICIENTS $\gamma\pm$ OF SOME COMMON ELECTROLYTES IN WATER AT 25°C

Mol kg^{-1}	0.001	0.005	0.01	0.05	0.1	0.5	1
HCl	0.965	0.929	0.905	0.830	0.794	0.757	0.809
H_2SO_4	0.830	0.639	0.544	0.340	0.265	0.154	0.130
NaOH	—	0.926	0.902	0.818	0.766	0.690	0.678
NaCl	0.965	0.927	0.902	0.819	0.744	0.681	0.657
KCl	0.965	0.927	0.902	0.816	0.770	0.649	0.607

(the Avogadro number) = $6 \cdot 10^{23}$ mol^{-1}, R (the gas constant) = 8.413 J $mol^{-1}K^{-1}$, ρ (the density of solution, in kg $m^{-3}\cdot 1000$), and ϵ (the permittivity of water) ≅ 78 · 8.85 pF m^{-1}. Hence, at T = 298 K, A = 0.509. For a binary electrolyte at I = 0.1, one may then expect a $\gamma_\pm$ of 0.69.

In practice, activity coefficients of ions can only be measured for ion pairs (e.g., a binary electrolyte) and as such are found in the appropriate critical tables. A selection of mean activity coefficients $\gamma_\pm$ is given in Table 2.

The simple expression in Equation 51 holds relatively well for low ionic strengths and deviates from reality at higher ones. Of the many extensions of the Debye-Hückel law (the so-called Limiting Law)[5] a useful formula is the one where two further constants are introduced so that

$$\log \gamma_\pm = A|z^2| \sqrt{I}/(1 + Ba\sqrt{I}) + CI \tag{53}$$

where $B = \sqrt{2 F^2/RT\epsilon}$ (numerically equal to $1.037 \cdot 10^8$ at 25°C), a is the effective diameter of the ion (0.3 to 0.5 nm), and C is an adaptable parameter, usually expressed as $-0.1z^2$. This formula predicts even the striking increases of activity coefficients at high concentrations of acids (e.g., $\gamma_{\pm(10\,M\,HCl)} = 10.44$; $\gamma_{\pm(10\,M\,NaOH)} = 3.23$).

REFERENCES

1. **Sörensen, S. P. L.,** Enzymstudien. II. Über die Messung und die Bedeutung der Wasserstoffkonzentration bei enzymatischen Prozessen, *Biochem. Z.*, 21, 131, 1909.
2. **Kolthoff, I. M.,** *Der Gebrauch von Farbindikatoren,* 4th ed., Berlin, 1932.
3. **Arrhenius, S.,** Einfluss der Neutralsalze auf die Reaktionsgeschwindigkeit der Verseifung von Aethylacetate, *Z. Phys. Chem.*, 1, 110, 1887.
4. **Brönsted, J. N.,** Acid and base catalysis, *Chem. Rev.*, 5, 232, 1928.
5. **Pearson, R. G.,** Hard and soft acids and bases, *J. Am. Chem. Soc.*, 85, 3533, 1963.
6. **Koryta, J. and Dvŏrák, J.,** *Principles of Electrochemistry,* John Wiley & Sons, Chichester, England, 1987.
7. **Debye, P. and Hückel, P.,** Lowering of freezing point and related phenomena, *Phys. Z.* 24, 185, 1923.

Chapter 2

pH IN BIOLOGICAL SYSTEMS

Arnošt Kotyk

TABLE OF CONTENTS

I. RELATIVE CONSTANCY OF INTRACELLULAR pH

In cellular and subcellular biology, which is a science describing phenomena proceeding in 99% of cases in an aqueous solution, pH can be considered as one of the essential intensive parameters, much like temperature or viscosity.

Every cell and organelle, and every type of body fluid is characterized by a narrow range of pH values within which it can only function properly.

The greatest number of measurements have been done on mammalian cells, particularly muscle and nerve cells and on erythrocytes.[1] With very few exceptions, the values obtained lie above pH 7 and, with no exception, are they greater than 6.81 which is the "neutrality" value at 37°C (see Chapter 1.I). It is only in plant cells, particularly their vacuoles, in the majority of microorganisms, and in some animal lysosomes, that pH_{in} values at around or below 6 are encountered.

The question why this is so is difficult to answer. However, it may follow from the fact that the principal cations of cells are K^+ (with Na^+ predominating in extracellular fluids), which means that conjugated to them are strong bases, and that the predominant anions are Cl^-, HCO_3^- and $H_2PO_4^-$, which are anions of weaker (with the exception of Cl^-) acids. Following the hydrolysis scheme (see Chapter 1.III.) the resulting pH tends to be slightly on the alkaline side. On the other hand, in plant and yeast vacuoles the anion is phosphate but the cation is often a basic amino acid, definitely a very weak base (Table 3).

The buffering power of the cell interior can be analyzed by using Equation 48a or 48b. If the total buffer concentration $c_T = c_B + c_{BH^+}$, and since Equation 44 is applicable, we can write

$$\beta = 2.3K_A c_{H^+} \cdot c_T/(K_A + c_{H^+})^2 \qquad (54)$$

Taking derivatives of β with respect to c_{H^+} it can easily be shown that the maximum buffering power is reached such that

$$\beta_{max} = 2.3c_T/4 = 0.58c_T \qquad (55)$$

If, like in a cell, the buffer contains a component that equilibrates with the surroundings, in this case CO_2, the situation is different. Production of intracellular acid shifts the equation

$$H^+ + HCO_3^- \rightleftharpoons H_2CO_3 \rightleftharpoons CO_2 + H_2O \qquad (56)$$

to the right and, as the CO_2 produced leaves the cell (while c_{CO_2} is constant), the shift toward the right is much more pronounced than the case would be in a closed system. Likewise, the addition of an alkaline solution shifts the equilibrium farther to the left.

The open-system buffering power due to CO_2 is then

$$\beta_{CO_2} = 2.3c_{HCO_3^-} \qquad (57)$$

It should be noted that this β_{CO_2} alone may amount to 30 m*M* in mammalian muscle cells and to considerably more in yeast cells.

In this context, an important observation should be made, as follows from considerations in this and the previous section, viz., that flushing a cell suspension or tissue with CO_2 will decrease not only the extracellular pH, but, owing to rapid permeation of CO_2 into cells ($P_{CO_2} \cong 0.5$ cm s^{-1}), also the intracellular pH. This was, for instance, made use of for bringing down the pH_{in} of sugarbeet cells to diminish their sucrase activity.[2]

The relatively constant intracellular pH (see, however, the section on regulation) is not

Table 3
DISSOCIATION CONSTANTS AND ISOELECTRIC POINTS OF NATURAL AMINO ACIDS

Amino acid	pK_1	pK_2	pK_3	pI
Glycine	2.34	9.60	—	5.97
Alanine	2.35	9.69	—	6.02
Valine	2.32	9.62	—	5.97
Leucine	2.36	9.60	—	6.02
Isoleucine	2.36	9.68	—	6.02
Serine	2.21	9.15	—	5.68
Threonine	2.63	10.43	—	6.53
Phenylalanine	1.83	9.13	—	5.48
Methionine	2.28	9.21	—	5.75
Tryptophan	2.38	9.39	—	5.89
Tyrosine	2.20	9.11	10.07(OH)	5.66
Aspartic acid	2.09	3.86 (β-COOH)	9.82 (NH_2)	2.97
Asparagine	2.02	8.08	—	5.05
Glutamic acid	2.19	4.25 (γ-COOH)	9.67 (NH_2)	3.22
Glutamine	2.17	9.13	—	5.65
Lysine	2.18	8.95	10.53 (ε-NH_2)	9.74
Hydroxylysine	2.13	8.62	9.67 (ε-NH_2)	9.15
Histidine	1.82	6.02 (imida-zol.)	9.17 (NH_2)	7.59
Arginine	2.17	9.04	12.48 (guanidin.)	10.76
Cysteine	1.71	8.33 (SH)	10.78 (NH)	5.1
Cystine	1.65	7.85	—	4.74
Proline	1.99	10.60 (=NH)	—	6.30

Note: The isoelectric or isoionic points (pI) of bifunctional amino acids are computed from pI = (pK_1 + pK_2)/2. Those of trifunctional ones, such as lysine or glutamic acid, are calculated from the two dissociation constants governing the dissociation and association, respectively, of the species carrying no net charge: thus, pK_2 and pK_3 in the case of lysine, and pK_1 and pK_2 in the case of glutamic acid. Only in the case of cysteine does the third pK also come into play. The pK values without description correspond to the α-carboxyl and α-NH_2 groups.

entirely due to the buffering action of cytosolic constituents. Cells have developed strategies to maintain this pH actively by metabolic and transport processes. Two cases may serve as examples.

First, in mammalian cells, the principal ion-exchanging pumps are those driven by the splitting of ATP at one or another of the membrane ATPases,[3] principally tha Na,K-ATPase present in all plasma membranes, the K,H-ATPase in gastric oxyntic cells, and the vacuolar-type H-ATPase of some organelles, such as lysosomes. A powerful H^+-extruding system is the electron-transport chain of inner mitochondrial or thylakoid membranes, where the local gradient of the H^+ ions is usually immediately utilized as a source of energy for ATP synthesis.

Of all these, only the gastric K,H-ATPase, a rather specialized enzyme, indeed, and the vacuolar-type ATPases are directly involved in H^+ shifts, and it appears that their role lies elsewhere than in maintaining cytosolic pH. In the case of the gastric ATPase it is to create an extremely low-pH milieu in the stomach; the lysosomal ATPases generate a similar, even if more moderate, pH decrease within these usually digestive organelles.

The cytosolic pH, which tends to be lowered by metabolic production of acids, is probably regulated by different transport systems, two possibilities having been explored:

1. Various energy-dependent systems that would exchange Cl^-_{in} for OH^-_{out} or $HCO^-_{3\,out}$, or K^+_{out} for H^+_{in}, but, although evidence for the involvement of Cl^- and HCO_3^- is strong,[4] all these mechanisms would necessarily draw on the cell's energy supply and, thus, could only function for limited periods of time and under stress conditions. They may operate to compensate for cytosolic alkalinization,[5] but not for its more naturally occurring acidification.[6]
2. A generally found compensatory mechanism is the one of Na^+/H^+ exchange, Na^+ ions being abundant extracellularly so that in principle, an exchange (probably on a specific carrier protein) can take place without any requirement for energy to do electrical work.[7,8]

In the second example the H^+ transport is affected by the plasma membrane ATPase of plants and yeasts. This extrusion of H^+ ions appears to be driven by "anaerobic" metabolism, particularly glycolysis, where acidity is increased as phosphorylated acids, as well as free acids (e.g., glyceric, acetic, succinic)[9,10] are formed at the full or partial expense of the free energy of dehydrogenation or, at the initial stages, of ATP or of polyphosphate splitting. It should be noted in this context that either ATP or polyphosphate splitting gives rise to additional H^+ ions, since two terminal phosphates are generated from a single one and one extra H^+ ion is thus dissociated.

It appears (even if argued *a posteriori*) that the intracellular pH of primeval cells that was generated by a "passive" hydrolysis of salts of strong bases with weak acids, represented one of the selection-pressure factors that determined the optimum pH of intracellular enzymes. These pH_{opt} values are now almost without exception in the very slightly alkaline range (Table 4). It should be observed that the only enzymes that have their pH_{opt} distinctly below 7 are extracellular, such as acid phosphatase (5.4), pepsin (1.8 to 2.0), *Bacillus subtilis* α-amylase (5.0 to 6.0), bacterial β-amylase (4.0 to 5.0), microbial β-glucosidase (4.4 to 5.0), α-mannosidase (4.0 to 5.0), papain (6.0 to 6.5), sialidase (5.0 to 5.1), etc.

II. CATALYTIC EFFECTS OF pH

A. Enzymic Reactions

The molecular basis of the existence of an optimum pH value for an enzyme is now generally believed to be the correct protonation of the enzyme-active site for it to be functional. If the active site is not protonated (at high pH) or if it is doubly or perhaps multiply protonated (at low pH), the enzyme does not work. A reasonably simple kinetic scheme, consisting of a nonprotonated enzyme E, a singly protonated enzyme EH, and a doubly protonated enzyme EH_2, can serve as a basis for a kinetic description of the reaction.

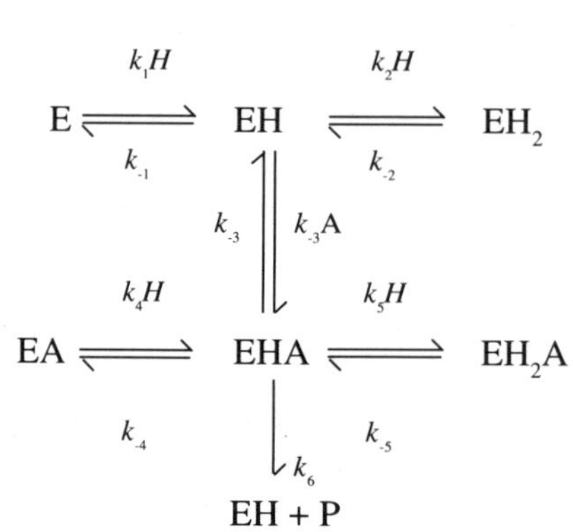

(I)

Table 4
OPTIMUM pH VALUES OF SELECTED ENZYMES

Enzyme	E.C. number	pH_{opt}
Alcohol dehydrogenase	1.1.1.1	8.4—9.5
L-Iditol dehydrogenase	1.1.1.14	7.4
Glyoxylate reductase	1.1.1.26	6.3—6.6
UDPglucose dehydrogenase	1.1.1.22	8.7
L-Lactate dehydrogenase	1.1.1.27	7.0—7.5
Malate dehydrogenase	1.1.1.37	7.4
Glucose-6-phosphate dehydrogenase	1.1.1.49	7.4—8.6
Xanthine oxidase	1.1.3.22	7.2—8.5
Glyceraldehyde-6-phosphate dehydrogenase	1.2.1.12	8.7
Alanine dehydrogenase	1.4.1.1	8.8—10.5
Glutamate dehydrogenase	1.4.1.2	7.6—8.6
L-Amino-acid oxidase	1.4.3.2	7.0—7.5
D-Amino-acid oxidase	1.4.3.3	9.0
Glutathione reductase	1.6.4.2	6.4—6.7
Urate oxidase	1.7.3.3	9.5
Catalase	1.11.1.6	7.0
Peroxidase	1.11.1.7	7.0
Lipoxygenase	1.13.11.12	9.0
Aspartate aminotransferase	2.6.1.1	7.6
Hexokinase	2.7.1.1	7.5
Pyruvate kinase	2.7.1.40	7.0—7.8
1-Phosphofructokinase	2.7.1.56	6.7—7.8
Acetate kinase	2.7.2.1	7.4
Phosphoglycerate kinase	2.7.2.3	6.8
Creatine kinase	2.7.3.2	9.0
Adenylate kinase	2.7.4.3	7.5
Nucleoside-diphosphate kinase	2.7.4.6	6.8—8.0
RNA-directed DNA polymerase	2.7.7.49	7.8—8.3
Triacylglycerol lipase	3.1.1.3	7.4
Phospholipase A_2	3.1.1.4	7.9—8.4
Acetylcholinesterase	3.1.1.7	7.0—7.2
Cholinesterase	3.1.1.8	8.2
Alkaline phosphatase (*E. coli*)	3.1.3.1	8.0
Glucose-6-phosphatase	3.1.3.9	6.0—7.0
Phospholipase C	3.1.4.3	6.6—8.0
Lecithinase D	3.1.4.4	5.6
Venom exonuclease	3.1.15.1	8.9—9.8
Spleen exonuclease	3.1.16.1	6.0
Ribonuclease T1	3.1.27.3	7.4
Ribonucleases (various)	3.1...	7.1—7.6
α-Amylase	3.2.1.1	7.5—7.8
α-Glucosidase	3.2.1.20	6.8
β-Galactosidase	3.2.1.23	7.2—7.7
β-Fructofuranosidase	3.2.1.26	4.7—6.2
Aminopeptidases (various)	3.4.11..	7.2—9.0
Dipeptidases (various)	3.4.13..	
	3.4.14..	7.6—9.0
	3.4.15..	
Carboxypeptidases (various)	3.4.16..	7.4—7.8
	3.4.17..	
Chymotrypsin	3.4.21.1	8.0
Trypsin	3.4.21.4	7.0—9.0

Table 4 (continued)
OPTIMUM pH VALUES OF SELECTED ENZYMES

Enzyme	E.C. number	pH_{opt}
Clostridium histolyticum collagenase	3.4.24.3	6.0—8.0
Asparaginase	3.5.1.1	7.2—8.4
Urease	3.5.1.5	8.0
Creatininase	3.5.2.10	8.0
Arginase	3.5.3.6	9.2—10.2
Adenosine deaminase	3.5.4.4	7.0—7.4
Fructose-biosphosphate aldolase	4.1.2.13	7.1—8.5
Citrate (*pro-3S*) lyase	4.1.3.6	8.0
Carbonate dehdyratase	4.2.1.1	6.0—10.0
Fumarate hydratase	4.2.1.2	7.8
Enolase	4.2.1.11	6.2—6.9
Phosphoglycerate mutase	5.4.2.1	7.0
Phosphoglucomutase	5.4.2.2	6.5—8.0

Here, A is the enzyme reaction substrate and P the product. The scheme can be analyzed for a steady state such that $dc_{EH}/dt = 0$, $dc_{EHA}/dt = 0$, etc., to obtain an expression in which it is useful (and probably permissible) to replace the ratios of rate constants. Thus: $k_{-1}/k_1 = K_1$, $k_{-2}/k_2 = K_2$, $k_{-4}/k_4 = K_{a_1}$, $k_{-5}/k_5 = K_{a_2}$, and $K_A \equiv K_m = (k_{-3} + k_6)/k_3$ while $v = k_6 c_{E_{total}}$. The resulting equation for the initial reaction rate is therefore

$$V = V'c_A/(c_A + K') \tag{58}$$

where

$$V' = V/(1 + c_{H^+}/K_{A_1} + K_{A_2}/c_{H^+}) \tag{59a}$$

and

$$K' = K_A/(1 + c_{H^+}/K_1 + K_2/c_{H^+}/(1 + c_{H^+}/K_{A_1} + K_{A_2}/c_{H^+}) \tag{59b}$$

whence

$$V'/K' = V/K_A(1 + c_{H^+}/K_1 + K_2/c_{H^+}) \tag{59c}$$

To determine the individual dissociation constants it is useful to plot log K', log V', and log(V'/K') against pH (Figure 4).

More complicated equations can be derived if one assumes that in scheme I the forms E and EH_2 can also associate with substrate A to EA and EH_2A, respectively, and/or if there is another intermediate form involved on the path from EH to EHA to EH + P, such as indeed occurs in many enzyme reactions.[11]

It is useful in this context to realize the detailed significance of the consecutive dissociation constants K_1 and K_2 which applies to any enzyme-substrate or receptor-ligand binding to multiple binding sites. In the case of two binding sites for protons the protonation may be envisaged in two ways,

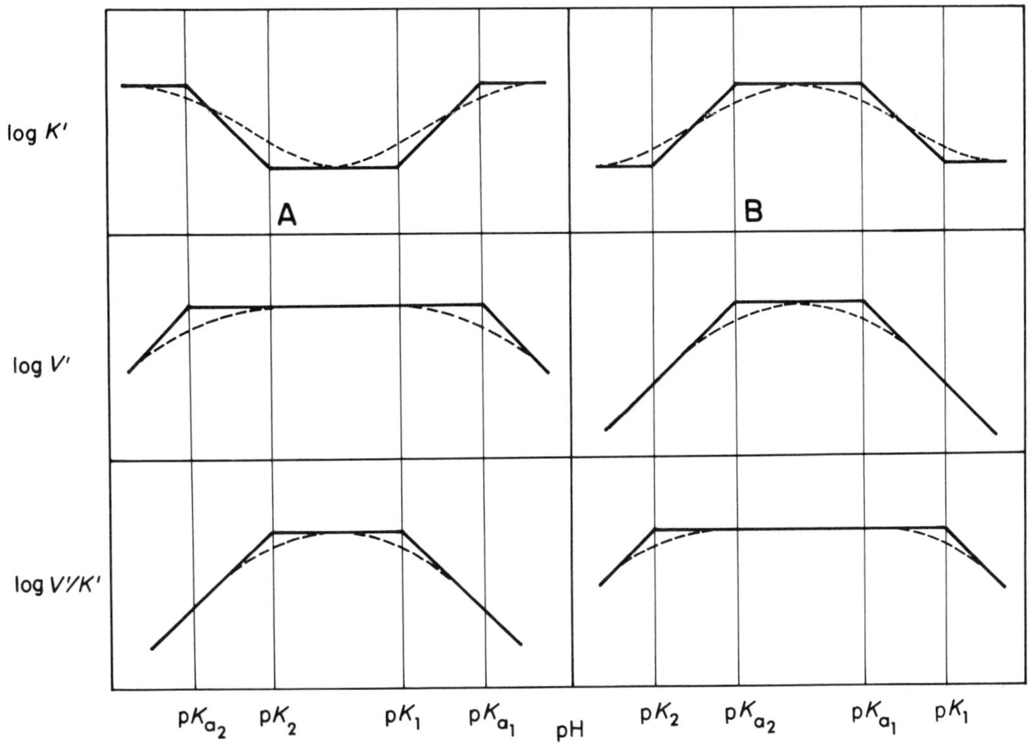

FIGURE 4. Different ways of plotting the pH dependence of an enzyme reaction, making it possible to define the individual dissociation constants of the enzyme-substrate complex (pK_a) and of the enzyme-H^+ complex (pK). Panel A shows the situation when $pK_2 > pK_{a2}$ and $pK_1 > pK_{a1}$, while in panel B the opposite relations hold.

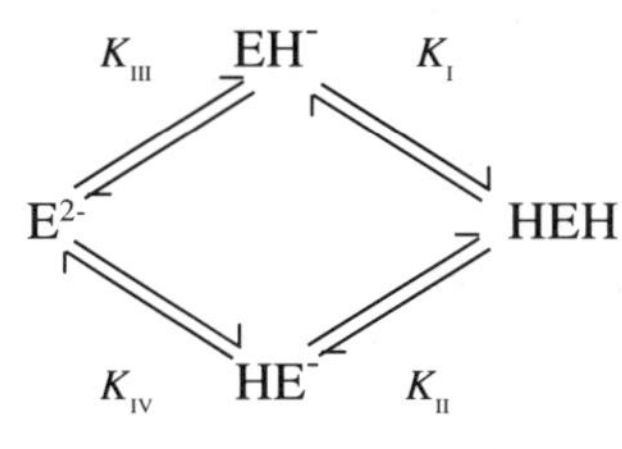

(II)

where $K_I - K_{IV}$ are the microscopic dissociation constants, defined as follows:

$$K_I = c_{EH^-} \cdot c_{H^+}/c_{HEH} \tag{60a}$$

$$K_{II} = c_{HE^-} \cdot c_{H^+}/c_{HEH} \tag{60b}$$

$$K_{III} = c_{E^{2-}} \cdot c_{H^+}/c_{EH^-} \tag{60c}$$

$$K_{IV} = c_{E^{2-}} \cdot c_{H^+}/c_{HE^-} \tag{60d}$$

while

$$K_1 = c_{H^+} \cdot (c_{EH^-} + c_{HE^-})/c_{HEH} \tag{61a}$$

and

$$K_2 = c_{H^+} \cdot c_{E^{2-}}/(c_{EH^-} + c_{HE^-}) \tag{61b}$$

If it is assumed that all the microscopic constants are intrinsically identical ($=K$), it can be shown that

$$K_1 = K_I + K_{II} = 2K \tag{62a}$$

and

$$1/K_2 = 1/K_{III} + 1/K_{IV} = 2/K \tag{62b}$$

Hence, necessarily

$$K_1 = 4K_2 \tag{63}$$

If there were n dissociable or binding sites it would hold that for the i-th site

$$K_{(i)} = Kn(n - i + 1)/i \tag{64}$$

This means that if the second macroscopic dissociation constant (here K_2) is found to be equal to one fourth of the first constant (here, K_1), the microscopic binding constants are, in fact, equal. In reality, K_1 is usually much greater than K_2. Thus, for oxalic acid, $K_1/K_2 = 920$, for malonic acid $K_1/K_2 = 650$, etc. However, in the enzyme-active site where two residues of the same amino acid may be involved in catalysis, $pK_2 - pK_1 = 0.6$ may easily be observed.

B.Transport Systems

Like enzymes, all specific transport systems are subject to catalytic pH effects, and pH optima of transport reactions are commonly reported in the literature, particularly for transport by unicellular organisms where the external pH can be varied without detrimental effects on the organism. In such cases, however, one should be aware that the carrier or channel binding site is alternately exposed to two solutions of possibly different pH. This may complicate matters somewhat as reflected in the corresponding rate equation,[12] analogous to Equation 58. J_{max} here is, in fact V', K_T is K'. Assuming that the values of K_1, K_2, K_{A_1}, and K_{A_2} at one membrane face are each identically equal to those at the other face, and introducing a as the rate constant of transmembrane movement of the EH complex and b as the rate constant for the EHA complex, we have

$$J_{max} = \frac{2c_{E_{total}}b}{1 + c_{H^+_{out}}/K_{A_2} + K_{A_1}/c_{H^+_{out}} + (b/a)(1 + c_{H^+_{in}}/K_{A_2} + K_1/c_{H^+_{in}})} \tag{65a}$$

$$K_T = \frac{K_A(2 + K_1/c_{H^+_{out}} + c_{H^+_{out}}/K_2 + K_1/c_{H^+_{in}} + c_{H^+_{in}}/K_2)}{1 + c_{H^+_{out}}/K_{A_2} + K_{A_1}/c_{H^+_{out}} + (b/a)(1 + c_{H^+_{in}}/K_{A_2} + K_1/c_{H^+_{in}})} \tag{65b}$$

In the diagnostic plot shown in Figure 4 the consequences of taking into account the intracellular pH are such as shown in Figure 5.

In addition to the above, there is an indirect kinetic effect of pH on membrane processes involving the transport of ions, either primary, where ions are driven by a chemical reaction, or secondary, where they serve as the driving species for transport, viz., the surface potential.

The surface charge on biomembranes, caused by an excess of negative (carboxyl and

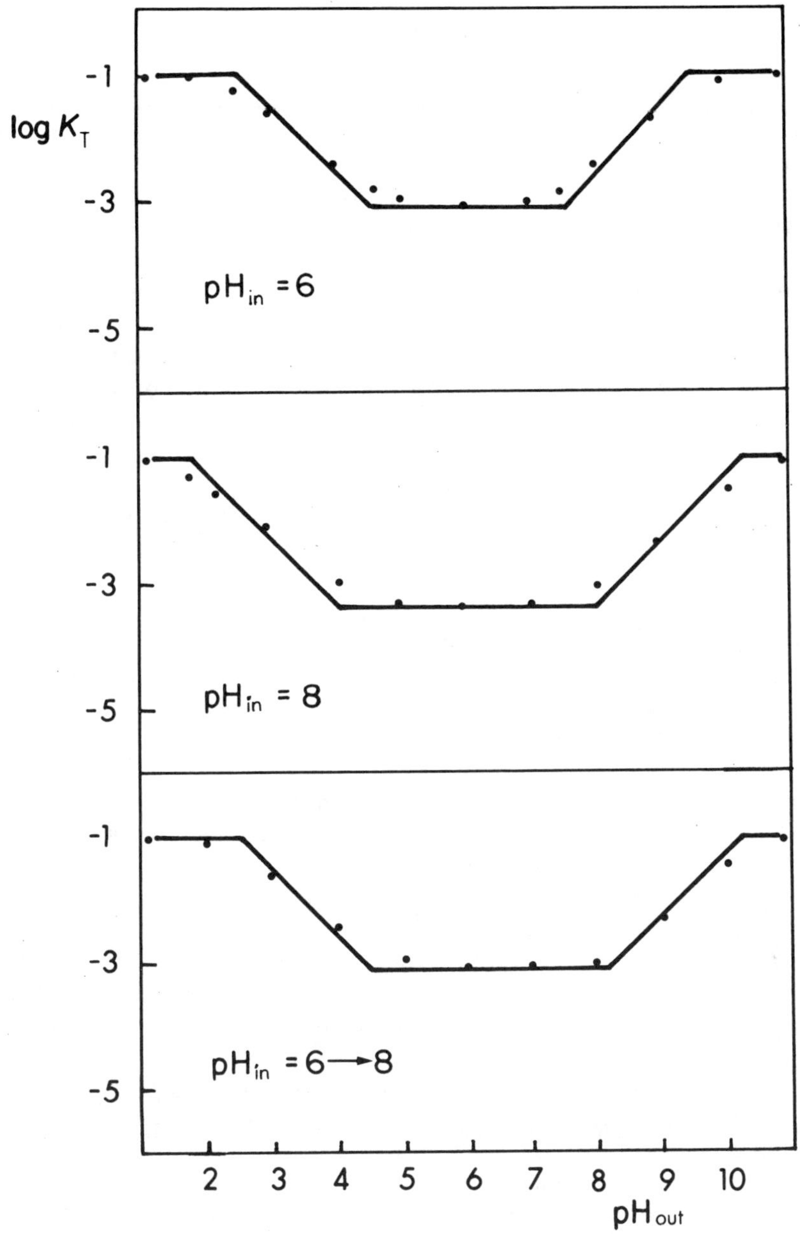

FIGURE 5. Dependence of the half-saturation constant of a transport reaction (K_T) on extracellular pH in the case that the intracellular pH is equal to 6 or 8, or to 6 at low and to 8 at high extracellular pH values.

phosphoryl) groups, is of the order of 1 to 100 mC m^{-2}, or 10^7 to 10^8 anionic groups per average animal cell. The surface charge, by attracting oppositely charged ions from the surrounding medium, generates, in the end, a double-layered distribution of charges around the cell, the potential formed by the charges nearest the membrane being called the surface potential.

This surface potential ψ_s results in changing the concentration of charged species near the membrane such that

$$c_{\mathrm{ion,s}} = c_{\mathrm{ion,b}}\, e^{-zF\psi_S/RT} \tag{66}$$

where the s and w subscripts stand for surface and bulk aqueous medium, respectively, and z is the number of positive charges on the ion. Clearly, both the half-saturation constant K_T and the maximum rate of transport J_{max} can only be determined on the basis of the bulk concentration and must be corrected according to Equation 66. Depending on whether an anion or a cation is transported and on whether a neutral solute or the driving cation binds first, the apparent transport parameters change quite substantially.

Now not only does the surface potential affect concentrations in membrane vicinity, but also artificially added ions alter the surface potential. Specifically, cations decrease its absolute value and do so depending on their valency which appears as the power exponent. Hydrogen ions do the same so that a very low pH not only does not support catalysis, but it also increases the K_T of the transport of cations, as well as of neutral solutes driven by cations.

III. REGULATORY EFFECTS OF INTRACELLULAR pH

In addition to the catalytic effects of pH exhibited by virtually all enzymes there exist fine controls of various cellular functions by pH_{in}. This view may seem to modify the statement made at the beginning of this chapter that pH_{in} is virtually constant. It is, indeed, relatively constant under given conditions, and it does not change, in general, if the external pH is altered. On the other hand, moderate changes of pH_{in} accompany various fundamental ontogenetic (long-term) as well as metabolic (short-term) phenomena.[13]

- Eggs of various animals (sea urchins have been studied most) become alkalinized after fertilization, probably due to extrusion of the so-called fertilization acid.[14] A similar effect has been observed with sperm upon dilution in sea water — an exchange of Na^+ for intracellular H^+ appears to be the underlying mechanism. The shifts of pH_{in} toward the alkaline side represent 0.4 to 0.5 U.[15]
- The cell cycle of unicellular objects ranging from the yeast *Saccharomyces cerevisiae*[16] to protozoans[17] to spleen lymphocytes[18] is accompanied by changes of pH_{in}. These changes are particularly pronounced in the unicellular eukaryotes, where pH_{in} shifts up by about 0.3 U. prior to DNA synthesis or mitosis.
- A remarkable effect of lowering pH_{in} is observed in high-density suspensions of fibroblasts,[19] which cease growing at a certain density. In all likelihood, this pH_{in} lowering is caused by CO_2 which permeates freely across cell membranes[20] and is produced in higher amounts at greater suspension densities. A similar phenomenon may underlie the inhibition of various metabolic processes (also by CO_2) observed in various microorganisms, especially yeasts.[21]
- A number of information-carrying compounds, such as insulin (in muscle cells), thrombin (in platelets), and peptide chemotactic factors (in neutrophils), bring about pH_{in} changes, generally an elevation by some 0.2 U.[22]

The actual target of the lowered or increased pH under the conditions described above is not at all clear, although, apparently, some key, highly pH-sensitive enzymes may be the principal effectors of the pH influence. It should be borne in mind, however, that any quantitation at the molecular level will be difficult before a true local pH at the affected enzyme is known (see the last chapter).[23-25]

IV. THERMODYNAMIC EFFECTS OF ΔpH

The actual concentration of H^+ ions is a potent catalytic factor in enzyme and transport reactions, but it may also play a role in driving some reactions where H^+ ions are directly involved as one of the substrates. As the equilibrium constant of the reaction is, under the given conditions, unchanged by substrate concentration, the equilibrium concentration of the products of such a reaction should be raised by lower pH.

While this is probably an uncommon phenomenon in biology, there is a corollary of it in membrane transport where the concentration of H^+ ions or, more precisely, the difference of H^+ concentrations across the membrane is an extremely important source of energy for moving solutes across membranes, both from the external environment inward and to and from organelles within the cell.

In the so-called secondary active transport systems[26] the transmembrane potential difference and the gradient of the driving ion (H^+ in microorganisms and in plants, Na^+ in animals) constitute what is called the electrochemical potential gradient, defined (for H^+ ions) as

$$\begin{aligned}\Delta\tilde{\mu}_{H^+} &= \tilde{\mu}_{in} - \tilde{\mu}_{out} \\ &= \tilde{\mu}^{\circ}_{H^+_{in}} - \tilde{\mu}^{\circ}_{H^+_{out}} + F\Delta\phi + RT\ln(c_{H^+_{in}}/c_{H^+_{out}}) \qquad (67)\end{aligned}$$

where $\Delta\phi$ is the transmembrane potential in V, the other symbols having been defined before. The ion- and potential-dependent terms $\tilde{\mu}^{\circ}$ are taken to be identical at both membrane sides so that

$$\begin{aligned}\Delta\tilde{\mu}_{H^+} &= F\Delta\phi + RT\ln(c_{H^+_{in}}/c_{H^+_{out}}) \\ &= 96490\ \Delta\phi - 5800\ \Delta\text{pH} \qquad (\text{in J mol}^{-1};\ \text{at } 30°\text{C}) \qquad (68)\end{aligned}$$

where $\Delta\text{pH} = \text{pH}_{in} - \text{pH}_{out}$.

This is the rigorous expression for the total energy stored in the gradient per mole, but it is now common usage to divide this with F and obtain the protonmotive force

$$\text{pmf} = \Delta\phi - 0.060\ \Delta\text{pH} \qquad (\text{in V};\ \text{at } 30°\text{C}) \qquad (69)$$

If the operation of the H^+-driven transport is obligatorily coupled with the dissipation of the $\Delta\tilde{\mu}_{H^+}$, it should hold that the free energy provided by $\Delta\tilde{\mu}_{H^+}$ $(= \Delta G_{H^+})$ should be equal in equilibrium to the free energy stored in the ratio of concentrations of the driven solute S $(= \Delta G_s)$ such that

$$n\ \Delta\tilde{\mu}_{H^+} = -\Delta\mu_{solute} \qquad (70)$$

where n is the number of hydrogen ions bound to the carrier. Then

$$RT\ln(c_{s_{in}}/c_{s_{out}}) = -n(F\ \Delta\phi + RT\ln[c_{H^+_{in}}/c_{H^+_{out}}]) \qquad (71)$$

This expression is valid for a neutral carrier. If the free carrier itself carries z positive charges, Equation 71 changes to

$$RT\ln(c_{s_{in}}/c_{s_{out}}) = -n(zF\ \Delta\phi + RT\ln[c_{H^+_{in}}/c_{H^+_{out}}]) \qquad (72)$$

This, then, is easily transformed into an expression relating directly the maximum accumulation ratio of the H^+-driven solute to the existing ΔpH and $\Delta\phi$. Thus,

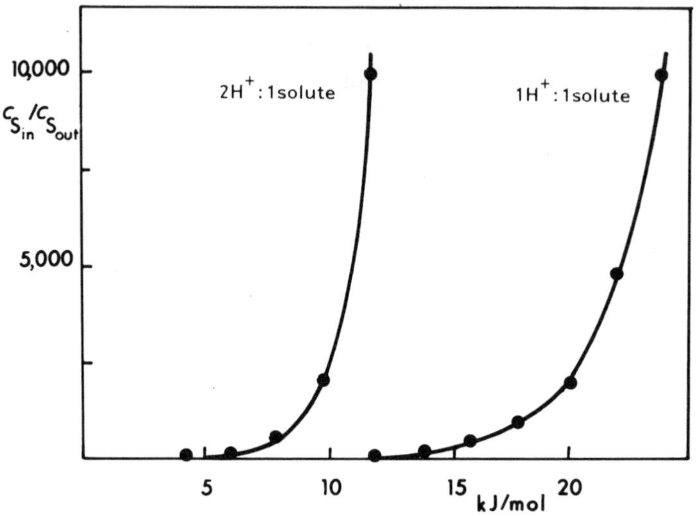

FIGURE 6. Theoretical prediction of maximum accumulation ratio of solute S driven by the electrochemical potential gradient of H^+ (in kJ mol^{-1}) for the case that 1 or 2 H^+ are carried across the membrane for each solute molecule.

$$c_{S_{in}}/c_{S_{out}} = (c_{H^+_{out}}/c_{H^+_{in}})^n \ e^{-(n+z)F\Delta\phi/RT} \tag{73}$$

Now a change of pH_{in} by one pH unit will affect the accumulation ratio of S by a factor of 10, even if $n = 1$. If $n = 2$, the change of the ratio would be by a factor of 100 (Figure 6).

Expression 73 is easily arrived at by a kinetic approach. Taking a simple model of a carrier C capable of binding solute S and driving ion H, such that only the C-H-S complex and the free carrier are mobile (i.e., they can expose themselves alternately to the one and to the other membrane face). (If the C-H complex were mobile the electrochemical potential gradient of H^+ ions would be wastefully dissipated. If the C-S complex were mobile it could not make use of the energy stored in the electrochemical potential gradient and would cause thermodynamic slipping of the whole system.)

The derivation proceeding from the assumption of a constant amount of carrier shows that both the ΔpH and the $\Delta\phi$ can exert an influence on the maximum rate of transport J_{max} and the half-saturation constant K_T. If the carrier system operates under rapid-equilibrium conditions, i.e., if the dissociation reactions of solute S and ion H proceed faster than the movement (or reorientation) of the carrier across the membrane (as they apparently mostly do), a lowering of external pH will quite generally lead to an increase of J_{max} and a decrease of K_T. If the intracellular pH is lowered, J_{max} will in most cases decrease while K_T will usually decrease for a neutral carrier but may under some modeling conditions increase for a positively charged carrier. An increase of the membrane potential difference will always increase J_{max} and decrease K_T. (Only for a positive carrier may the effect be either up or down.) These effects can be used as diagnostic features showing under what particular mechanism the given carrier operates.

If, however, the rapid-equilibrium assumption is not made, i.e., if the transmembrane movement is of similar rate as the membrane surface dissociation, the rate constants may be so manipulated that all kinds of response to ΔpH changes can be expected.

Clearly, this is a disconcerting situation with all H^+-driven transports which, in common with other specific membrane transports, behave like enzymic reactions catalytically affected by the ambient pH — as discussed in Section II of this chapter. It is particularly at very low and very high pH values that both the general protonation of the carrier and the amount of "thermodynamic" binding of H^+ ions can affect the values observed.

Then, again, it appears in most cases that for a proton-driven system to operate properly the primary active transport generating the $\Delta\tilde{\mu}_{H+}$ must function simultaneously. If it is blocked by a (specific) inhibitor, transport of secondarily driven solutes stops, although an artificial pH difference should be able to drive the transport easily enough. Here, simply the exponent of Equation 73 is equal to zero, making the equation into

$$c_{s_{in}}/c_{s_{out}} = (c_{H^+_{out}}/c_{H^+_{in}})^n \tag{74}$$

V. SIGNIFICANCE OF pH IN MICROSCOPIC OBJECTS

Substitution of concrete intracellular pH values into Equations 54 and 58 leads to precise but possibly incorrect values for two related reasons:

1. The cell pH is distinctly heterogeneous due to the presence of different organelles, nonuniformly distributed adsorption surfaces, and active movement of H^+ ions across various cell membranes, with the concomitant generation of pH gradients near the membrane surfaces. This question will be dealt with in some detail in the last chapter of the book.
2. It is often argued that in the microscopic dimensions it may not be justified to speak of H^+ ion concentrations. True enough, in a mitochondrion with a volume of some 0.5 μm^3, there may be only some ten H^+ ions present at any time at the physiological pH, while in a *Micrococcus* there may be, in fact, less than one H^+ ion per cell.

However, this view is not correct because the oxonium ion activity is a statistical term only. By a rapid dissociation of a proton donor inside the organelle, oxonium ions are formed, but subsequently disappear in a very rapid reaction with a conjugate proton acceptor. The time average seems to indicate that a fraction of an oxonium ion is present in the solution. In fact, the acidity of a solution is determined by the concentration of proton donors and proton acceptors which are represented inside the organelle by sufficiently large numbers of particles. The proton donors and acceptors are either components of a buffer system, but, in the extreme case of a completely unbuffered solution, the water molecules themselves act both as proton donors and proton acceptors.

REFERENCES

1. **Roos, A. and Boron, W. F.,** Intracellular pH, *Physiol. Rev.*, 61, 296, 1981.
2. **Bohuslavská, M., Zahradníček, J., Michaljaničová, D., and Kotyk, A.,** Effects of chemical and physicochemical factors on sugar-beet invertase activity, *Biol. Plant.*, 26, 303, 1984.
3. **Carafoli, E. and Scarpa, A.,** Transport ATPases, *Ann. N.Y. Acad. Sci.*, p. 402, 1982.
4. **Russell, J. M. and Boron, W. F.,** Role of chloride transport in regulation of intracellular pH, *Nature*, 264, 73, 1976.
5. **Vaughan-Jones, R. D.,** Chloride activity and its control in skeletal and cardiac muscle, *Philos. Trans. R. Soc. London Ser. B*, 299, 537, 1982.
6. **Vanheel, B., de Hemptinne, A., and Leusen, I.,** Analysis of Cl-HCO_3 exchange during recovery from intracellular acidosis in sheep cardiac Purkinje strands, *Am. J. Physiol.*, 246, C391, 1984.
7. **Piwnica-Worms, D., Jacob, R., Shigeto, N., Harres, C. R., and Lieberman, M.,** Na/H exchange in cultured chick heart cells. Secondary stimulation of electrogenic transport during recovery from intracellular acidosis, *J. Mol. Cell. Cardiol.*, 18, 1109, 1986.

8. **Moolenaar, W. H.,** Regulation of cytoplasmic pH by Na^+/H^+ exchange, *Trends Biochem. Sci.,* 11, 141, 1986.
9. **Sigler, K., Knotková, A., and Kotyk, A.,** Factors governing substrate-induced generation and extrusion of protons in the yeast *Saccharomyces cerevisiae, Biochim. Biophys. Acta,* 643, 572, 1981.
10. **Sigler, K., Kotyk, A., Knotková, A., and Opekarová, M.,** Processes involved in the creation of buffering capacity and in substrate-induced proton extrusion in the yeast Saccharomyces cerevisiae, *Biochim. Biophys. Acta,* 643, 583, 1981.
11. **Laidler, K. J. and Bunting, P. S.,** *The Chemical Kinetics of Enzyme Action,* Clarendon Press, Oxford, 1973.
12. **Kotyk, A.,** pH effects on membrane transport, in *Structure and Properties of Cell Membranes,* Vol. 2, Benga, Gh., Ed., CRC Press, Boca Raton, FL, 1985, 11.
13. **Busa, W. B. and Nuccitelli, R.,** Metabolic regulation via intracellular pH, *Am. J. Physiol.,* 246, R409, 1984.
14. **Runnstrom, J.,** Zur Kenntnis der Stoffwechselvorgänge bei der Entwicklungserregung des Seeigeleies, *Biochem. Z.,* 258, 257, 1933.
15. **Johnson, C. H., Clapper, D. L., Winkler, M. M., Lee, H. C., and Epel, D.,** A volatile inhibitor immobilizes sea urchin sperm in semen by depressing the intracellular pH, *Dev. Biol.,* 98, 493, 1983.
16. **Gillies, R. J., Ugurbil, K., den Hollander, J. A., and Shulman, R. G.,** ^{31}P NMR studies of intracellular pH and phosphate metabolism during cell division cycle of Saccharomyces cerevisiae, *Proc. Natl. Acad. Sci. U.S.A.,* 78, 2125, 1981.
17. **Gillies, R. J. and Deamer, D. W.,** Intracellular pH changes during the cell cycle in Tetrahymena, *J. Cell Physiol.,* 100, 23, 1979.
18. **Gerson, D. F., Kiefer, H., and Eufe, W.,** Intracellular pH of mitogen-stimulated lymphocytes, *Science,* 216, 1009, 1982.
19. **Ceccarini, C. and Eagle, H.,** pH as a determinant of cellular growth and contact inhibition. *Proc. Natl. Acad. Sci. U.S.A.,* 68, 229, 1971.
20. **Jones, R. P. and Greenfield, P. F.,** Effect of carbon dioxide on yeast growth and fermentation, *Enzyme Microbiol. Technol.,* 4, 210, 1982.
21. **Kotyk, A. and Michaljaničová, D.,** Suspension density and accumulation ratio of sugars and amino acids in yeasts, *Folia Microbiol.,* 32, 459, 1987.
22. **Molski, T. F. P., Naccache, P. H., Volpi, M., Wolpert, L. M., and Sha'afi, R. I.,** Specific modulation of the intracellular pH of rabbit neutrophils by chemotactic factors, *Biochem. Biophys. Res. Commun.,* 94, 508, 1980.
23. **Padan, E. and Schuldiner, S.,** Intracellular pH regulation in bacterial cells, *Methods Enzymol.,* 125, 337, 1986.
24. **Schuldiner, S., Agmon, V., Brandsma, J., Cohen, A., Friedman, E., and Padan, E.,** Induction of SOS function by akaline intracellular pH in *Escherichia coli, J. Bacteriol.,* 168, 936, 1986.
25. **Zavoico, G. B., Cragoe, E. J., Jr., and Feinstein, M. B.,** *J. Biol. Chem.,* 261, 13160, 1986.
26. **Kotyk, A.,** Coupling of secondary active transport with $\Delta\tilde{\mu}_{H+}$, *J. Bioenerg. Biomembr.,* 15, 307, 1983.

Chapter 3

METHODS OF DETERMINING INTRACELLULAR pH WITH ELECTRODES

Arnošt Kotyk

TABLE OF CONTENTS

I. INTRODUCTION

Over the past 40 years a fairly wide variety of techniques have been employed to assay the intracellular pH, and until quite recently there was little effort to obtain more than an average value for the whole cell. It was only several years ago that advanced techniques of NMR spectroscopy and of dual-excitation fluorescence shed more light on the local pH variation within the cell.

It should be said at the very outset of this chapter that there is no practicable way of direct measurement of the concentration of H^+ inside cells. The theoretical possibilities of distinguishing H^+ by infrared (IR) or Raman spectroscopy (H_3O^+, $H_5O_2{}^+$, etc.) or by ^1H NMR have no practical value, as none of these approaches can monitor 0.1 μM concentrations of H^+ as correspond to pH 7, against a nearly 100 *M* background signal of rapidly exchanging protons. The figure arises from the fact that water is 55 *M*, hence, 110 *M* hydrogen atoms. If an appropriate correction is taken for the amount of intracellular water and for the amount of H in the dry matter, a value of about 100 *M* is obtained.

Thus, all intracellular pH measurements are indirect, relying upon a pH indicator that reflects the ambient pH. The indicator may be represented by a pH-sensitive electrode or by a pH-sensitive chemical substance.

In this chapter we shall deal with the pH electrode techniques. Two approaches can be used here. Direct measurement *in situ* requires special microelectrodes and sensitive micromanipulation to introduce them into cells. Measurement of the pH resulting from permeabilizing the cells, either by mechanical rupture or by making sufficiently large pores in the plasma membrane to permit complete leakage of the cell interior into the external medium is less direct.

II. INTRACELLULAR MICROELECTRODES

Conceptually most straightforward, but technically very exacting, is the method using pH-sensitive electrodes small enough to penetrate into single cells without damaging them. This means in practice that the portion of the electrode that penetrates the cell should have a diameter not exceeding 1 μm. Only with giant cells, such as some nerve axons, can the size be considerably larger.

Most of these microelectrodes are simply miniature versions of the long-known and long-used glass electrode made of a "pH sensitive" glass. The surface of the 1- to 10-μm-thick electrode glass is believed to generate a potential, depending on the concentration of H^+ ions, caused by their permeation through the thin glass membrane. The liquid-junction potential across the membrane is thus given by

$$E = (RT/F)\ln(c_{H^+_{in}}/c_{H^+_{out}}) \tag{75}$$

As the inside of the electrode is maintained at a low and constant pH (HCl or an acid buffer), the potential becomes a straightforward function of the pH outside the electrode

$$E = \text{constant} + 2.3 \text{ pH } RT/F \tag{76}$$

From this it directly follows that at 25°C, a pH change of 1 corresponds to a potential change of almost exactly 60 mV.

The design of such a microelectrode dates back some 30 years[1] and underwent several major innovations in its development. The first routinely usable glass microelectrode was developed in 1959,[2] but was made of sodium- or potassium-sensitive glass. This was the protruding-tip type (Figure 7A). Thomas[3,4] developed a superior type of such electrode where

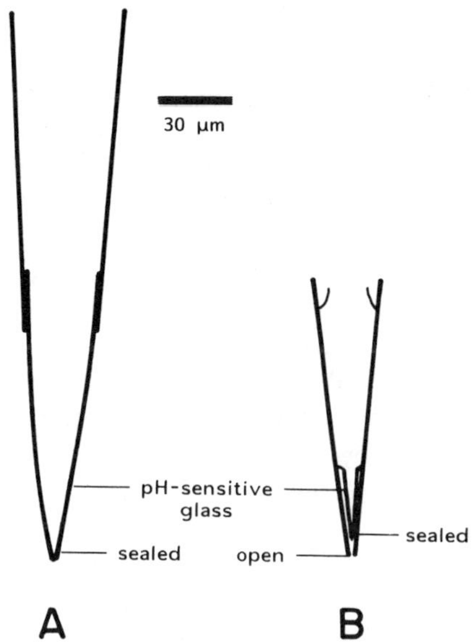

FIGURE 7. Drawing of the tips of the protruding-tip microelectrode (A) and of the recessed-tip microelectrode (B) showing that the recessed-tip modification is smaller and, hence, less damaging to the impaled cell.

the sensitive tip was recessed. This made the electrode more robust and, at the same time, the part introduced into the cell could be made at a smaller diameter (Figure 7B).

A. A Recessed-Tip Microelectrode

Excerpts from Thomas' work[4] will be used to describe the preparation of a recessed-tip microelectrode.

A commercial microforge attached to a triple micromanipulator is needed, but a ''designer's'' type may be finer and easier to control.

The micropipette body of the future microelectrode must be welded together from two types of glass: an insulating coat made of 2-mm tubing (Corning 1720® or Schott Supremax®) and a pH-sensitive inner tip made of 1-mm tubing (NaCaS 22-6 which is Corning 0150®).

Both types of tubing are first pulled out into slowly tapering, thin-tipped micropipettes, using a micropipette puller (e.g., Kopf 700C® or Ensor®) and a good deal of skill to get the right angle of tapering at the tip and the correct diameter at its opening, is required.

The inner micropipettes are first selected according to their diameter at a given distance from the tip and then sealed with a microforge under visual supervision through a horizontally mounted objective (10×, 25×, or 40×). The heating and sealing is effected with a rhodium-platinum coil and should be done gently, just up to the beginning of melting and slight distortion. The sealed part of the tip should not be longer than 5 μm.

The outer, insulating micropipette is best prepared in two stages of heating. In the first stage it should be carefully pulled out at the heated site by about 1 cm, then cooled and repositioned. In the second stage very gentle heating should be applied to pull it out completely (i.e., until it breaks apart).

The subsequent matching of the two pipettes together is a demanding and often vexing job. The inner pipette must be inserted into the outer one not only to allow for sufficient area of contact between them to make subsequent sealing easier, but also to ensure the exact separation of the inner, sealed tip from the outer, tapering but open tip (cf. Figure 7B).

The procedure used by Thomas was to fix the outer micropipette in the microforge, and then to attach the inner micropipette to the end of a piece of 0.5-mm steel tubing using sealing wax, and finally to connect the other end of the steel tubing to a cylinder with nitrogen or CO_2 (argon would do equally well). The gas should be allowed to reach a pressure of some 5 MPa (about 50 atmospheres), and the inner tubing should now be carefully lowered into the outer one. Provided a satisfactory area of contact is obtained, heat should be applied from a platinum loop positioned exactly around the area of contact.

If a tight seal has been achieved (visual inspection will show this), the inner micropipette is trimmed and may be immediately filled with 100 m*M* NaCl (Na_3-citrate titrated to pH 6 with HCl). Bubbles should be removed with a rugged hair or whisker, and the electrode should be kept overnight at 70°C in a drying oven. Then an Ag/AgCl wire is sealed tightly to the top of the microelectrode, and the tip is dipped for some time in a chromic cleaning solution to remove any possible contamination. The final electrode should have a resistance in the vicinity of $10^{11}\Omega$. It it is too low, this indicates that the seal is leaky and the electrode is worthless. If the resistance is well above $10^{11}\Omega$, aging of the electrode might help to bring it down.

A slow response of the microelectrode can be remedied by enlarging the opening of the outer micropipette and/or soaking it briefly in 4 *M* NaOH with 0.1 *M* EDTA.

For the measurement itself the Ag/AgCl wire and the lead from an external reference electrode are connected to an electronic potentiometer, and the potential between them is recorded.

B. Other Types of Glass Microelectrodes

Several other versions of the fine-tipped microelectrode have been reported in the literature, all of them variations on the above-described theme.[5-8]

A truly new design appeared only with the double-barreled glass microelectrode.[9] This novelty was prompted by the fact that the potential measured with a pH-sensitive electrode is a sum of the actual pH response and the membrane potential of the impaled cell. This difficulty can be overcome in three ways:

1. The membrane potential can be estimated by another, independent measurement, using a KCl-filled microelectrode (plus a reference electrode in the external medium).
2. It can be measured in parallel in the same cell provided the cell is large enough not to be damaged by simultaneous insertion of the two microelectrodes (each connected to a reference electrode in the external solution).to a reference electrode in the external solution).
3. A dual electrode can be constructed, one compartment of which measures the membrane potential (a KCl-filled part), another compartment the membrane potential plus pH (a NaCl-filled part), each of the compartments being connected via an Ag/AgCl wire to an external electrode in the outer medium, and to a suitable measuring device.

The double-barreled glass electrode has proved its worth in a number of experiments, not only in isolated, perfused tissue, but also *in vivo*.[10]

The original procedure for constructing the double-barreled electrode[9] was to fix a length of 2mm borosilicate tubing with another (1 mm in diameter), to "wrap" the thinner tubing (under heating with a long heater coil) around the thicker one (this was kept in position by inserting a long steel needle through it), and, after removing the needle, pulling the twin tubes into a long, sharp tip (Figure 8). Subsequently, the straight thicker tube served as the sheath for a thin, pH-sensitive glass micropipette which was sealed into it by a procedure analogous to that described above for the single barrel recessed-tip electrode.

An innovative procedure[10] does away with the twisting. A complete single-barrel microelectrode is inserted into another, somewhat wider borosilicate micropipette and carefully glued with a two-component resin. However, a space is left at one place between the two

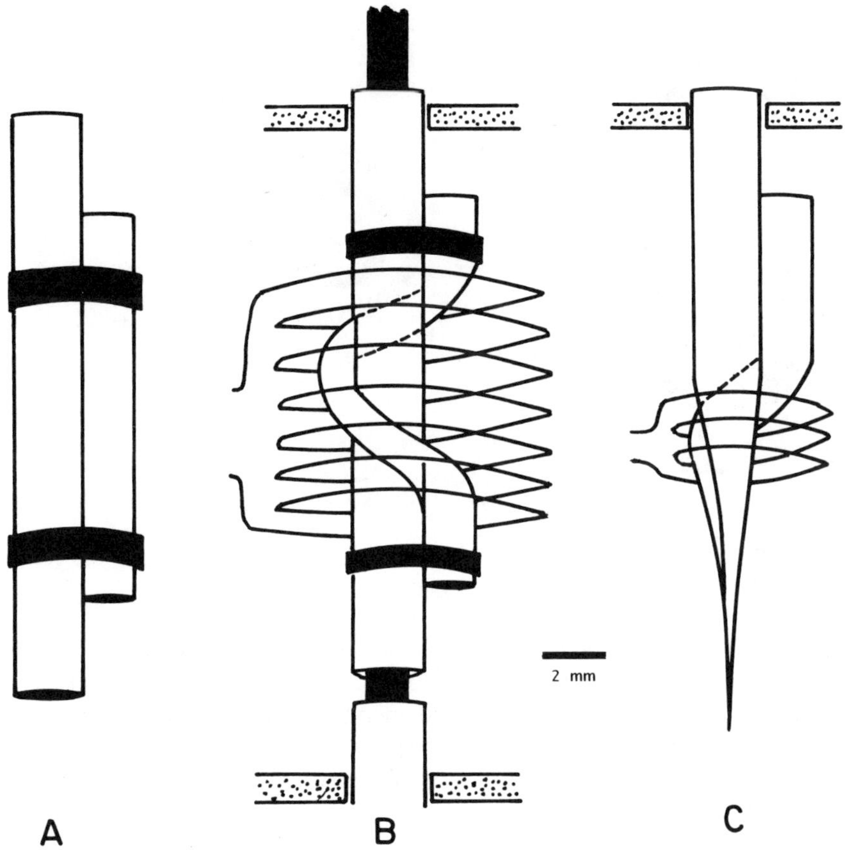

FIGURE 8. Drawing showing the process of preparation of a double-barreled micropipette as described in the text.

capillaries for subsequent insertion of the Ag/AgCl wire. Viewed from the cell to be impaled the two double-barreled electrodes look as shown in Figure 9. The inner pH electrode is again filled with a solution of 0.1 *M* NaCl and citrate buffered to pH 6; the outer, electrical potential electrode is filled with saturated K_2SO_4 containing 50 m*M* KCl. Both solutions are connected through Ag/AgCl wires to preamplifiers with input impedances of 1 TΩ (for pH) and 10 GΩ (for E_m).

Remarkably good results were obtained with this method in nerve fibers and muscle cells even *in vivo* (Figure 10).

C. Non-Glass Microelectrodes

Besides using glass pH-sensitive electrodes, attempts were made in the past to use an antimony electrode where the antimony tip is covered with a thin layer of Sb(III) oxide. This undergoes a fairly reversible reduction to metal antimony

$$Sb_3O_2 + 6H^+ + 6e^- \rightleftharpoons 2Sb + 3H_2O$$

the reaction serving as pH sensor.

Although some measurements were done with such an electrode (of the double-barrel type) quite recently,[11] their quality is definitely inferior to that of a good glass microelectrode.

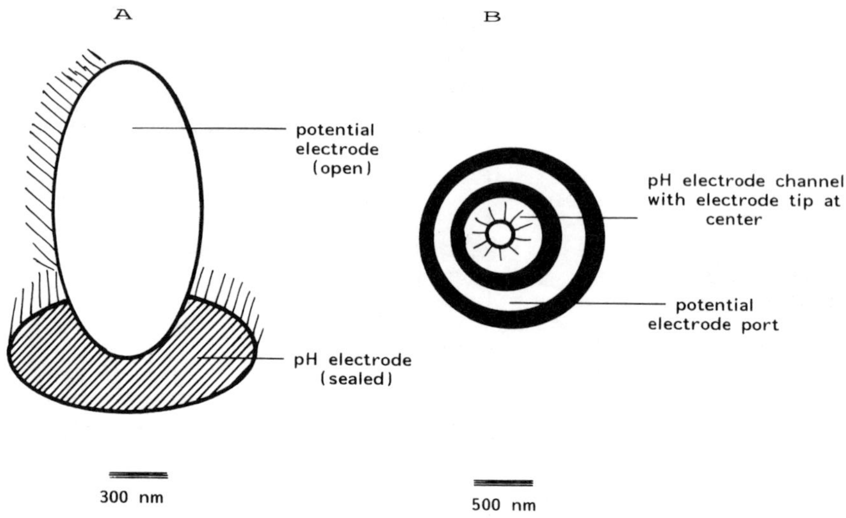

FIGURE 9. Schematic views of the tips of double-barreled electrodes prepared by the "twisting" procedure (A) and by the "insertion" procedure (B).

Attempts to use tungsten microelectrodes have not been reproducibly successful,[12] and, hence, they are not recommended for general use. An innovation in the field was the introduction of hydrogen ion-selective liquid membrane microelectrodes that are highly selective and easy to make. Moreover, their response is faster than that of glass microelectrodes.[13] The microelectrode is usually of the double-barreled type and contains a liquid ion exchanger plus nigericin.[14]

III. CELL PERMEABILIZATION

Perhaps the oldest technique for assessing intracellular pH was the disruption of cells and measurement of the attendant change of pH of the external solution when, with exact knowledge of intracellular aqueous volume, external volume, and the buffering capacity, it should be possible to calculate the original intracellular pH.

A number of cell-disruption techniques have been used: agitation with Ballotini beads,[15] repeated freezing and thawing,[16] sharp heating,[17] application of toluene,[18] and various detergents. Permeabilization has also been achieved with nystatin,[19,20] dimethyl sulfoxide,[21] and, most recently, with Sendai virus.[22]

There is no doubt that the measurement of pH in the resulting mixture is the simplest procedure possible. However, several objections have been raised against the use of this method. Not only can the permeabilization agents directly affect the pH, but disruption of cells can be accompanied by destruction of organelles and consequent leaking of their contents. Likewise, all the active membrane processes maintaining intracellular pH cease, while those taking place in the cytosol, such as glycolytic production of lactic acid, continue (unless the cells have been disrupted by drastic heating).

It is the author's opinion that the most serious objection is that what leaks out from the cells upon permeabilization is a buffer of unknown buffering power so that, even if pH changes of the original external medium can be calibrated by adding known amounts of a base or an acid, the pH change brought on by the intracellular content is not commensurate with them.

Last but not least, the method permits a single pH estimation from a sizable sample of cells which, moreover, are completely destroyed by the procedure.

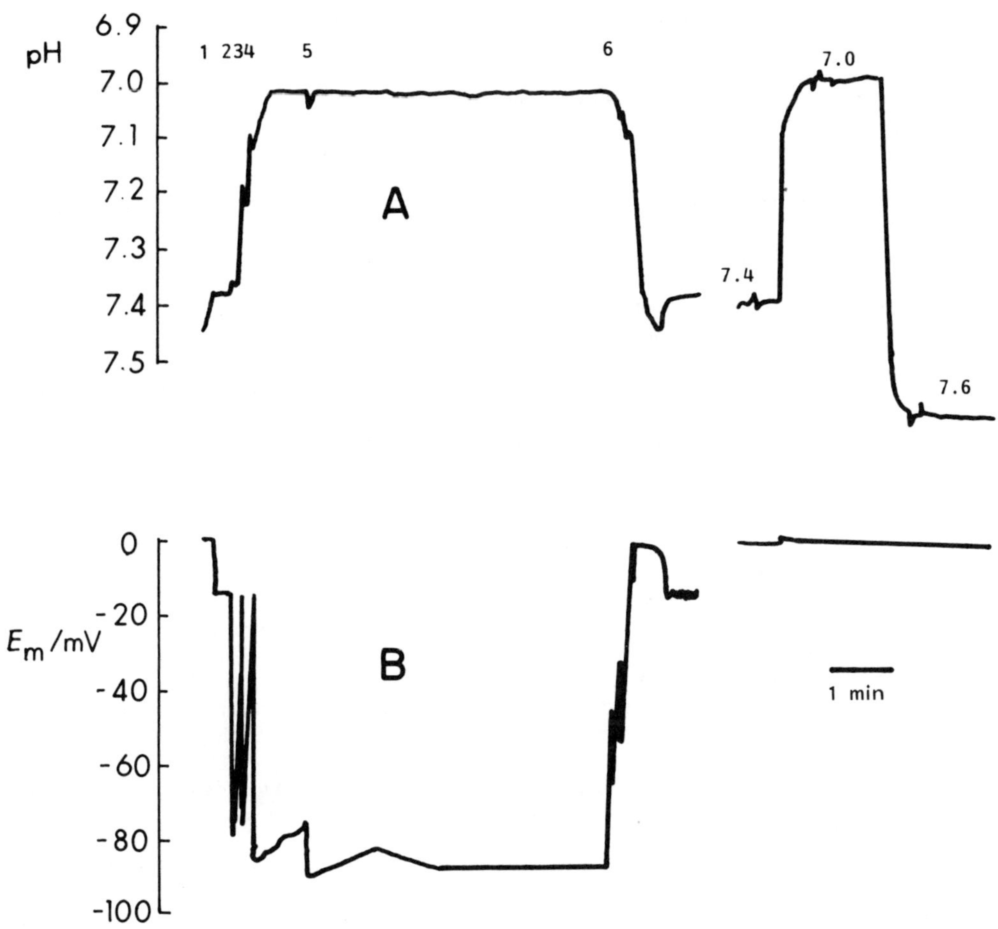

FIGURE 10. Recording of intracellular pH and of the membrane potential (E_m) using a concentric double-barreled microelectrode. The electrode was brought into contact with the gastrocnemius muscle at 1 and then impaled through several muscle fibers at 2, 3, and 4 until a relatively stable recording was obtained at 5. At 6, the microelectrode was withdrawn. The right-hand part of the graph shows calibration of the device using pH values of 7.4, 7.0, and 7.6. (Redrawn from Hagberg, H., Larsson, S., and Haljamäe, H., *Acta Physiol. Scand.*, 118, 149, 1983. With permission.)

Still, the method apparently yields satisfactory results with mammalian erythrocytes and, in this context, underwent several refinements, such as breaking the cells directly in a double-barreled capillary.[23]

REFERENCES

1. **Caldwell, P. C.,** An investigation of intracellular pH of crab fibres by means of micro-glass and micro-tungsten electrodes, *J. Physiol. (London),* 126, 169, 1954.
2. **Hinke, J. A. M.,** Glass microelectrodes for measuring intracellular activities of sodium and potassium, *Nature (London),* 184, 1257, 1959.
3. **Thomas, R. C.,** New design for sodium-sensitive glass electrode, *J. Physiol. (London),* 210, 828, 1970.
4. **Thomas, R. C.,** *Ion-Sensitive Intracellular Microelectrodes,* Academic Press, London, 1978.

5. **Kostyuk, P. G. and Sorokina, Z. A.,** On the mechanism of hydrogen ion distribution between cell protoplasm and the medium, in *Membrane Transport and Metabolism,* Kleinzeller, A. and Kotyk, A., Eds., Academic Press, London, 1961, 193.
6. **Lavallée, M.,** Intracellular pH of rat atrial muscle fibers measured by glass micropipette electrodes, *Circ. Res.,* 15, 185, 1964.
7. **Carter, N. W., Rector, F. C., Champion, R. T., and Seldin, D. W.,** Measurement of intracellular pH of skeletal muscle with pH-sensitive glass microelectrodes, *J. Clin. Invest.,* 46, 920, 1967.
8. **Pucacco, L. R. and Carter, N. W.,** A glass-membrane pH microelectrode, *Anal. Biochem.,* 73, 501, 1976.
9. **de Hemptinne, A.,** A double-barrel pH micro-electrode for intracellular use, *J. Physiol. (London),* 295, 5P, 1979.
10. **Hagberg, H., Larsson, S., and Haljamäe, H.,** A new design of double-barrelled microelectrodes for intracellular pH measurement in vivo, *Acta Physiol. Scand.,* 118, 149, 1983.
11. **Matsumura, Y., Kajino, K., and Fujimoto, M.,** Measurement of intracellular pH of bullfrog skeletal muscle and renal tubular cells with double-barreled antimony microelectrodes, *Membr. Biochem.,* 3, 99, 1980.
12. **Caldwell, P. C.,** An investigation of the intracellular pH of crab muscle fibres by means of micro-glass and micro-tungsten electrodes, *J. Physiol. (London),* 128, 169, 1954.
13. **Duffey, M. E.,** Intracellular pH and bicarbonate activities in rabbit colon, *Am. J. Physiol.,* 15, C558, 1984.
14. **Matsumura, Y. S., Aoki, S., Kajino, K., and Fujimoto, M.,** The double-barreled microelectrode for the measurement of intracellular pH, using liquid ion-exchanger, and its biological application, in Proc. Intl. Congr. Physiological Science, Budapest, 1980, 572.
15. **Lamanna, C. and Mallette, M. F.,** Use of glass beads for the mechanical rupture of microorganisms in concentrated suspensions, *J. Bacteriol.,* 67, 503, 1954.
16. **Conway, E. J. and Downey, M.,** pH values of the yeast cell, *Biochem. J.,* 47, 356, 1950.
17. **Conway, E. J., Brady, T. G., and Carton, E.,** Biological production of acid and alkali. II. A redox theory for the process in yeast with application to the production of gastric acidity, *Biochem. J.,* 47, 369, 1950.
18. **Gear, A. R. L., Rossi, C. S., Reynafarje, B., and Lehninger, A. L.,** Acid-base exchanges in mitochondria and suspending medium during respiration-linked accumulation of bivalent cations, *J. Biol. Chem.,* 249, 3403, 1967.
19. **Scholz, R., Schmitz, H., Bücher, T., and Lampen, J. O.,** Über die Wirkung von Nystatin auf Bäckerhefe, *Biochem. Z.,* 331, 71, 1959.
20. **Rybová, R., Janáček, K., and Kleinová, M.,** Properties of the proton transporting system in Hydrodictyon reticulatum, in *Plant Membrane Transport: Current Conceptual Issues,* Spanswick, R. M., Lucas, W. J., and Dainty, J., Eds., Elsevier/North-Holland, Amsterdam, 1980.
21. **Adams, B. G.,** Method for decryptification of α-glucosidase in yeast with dimethyl sulfoxide, *Anal. Biochem.,* 45, 137, 1972.
22. **Bashford, C. L., Alder, G., Micklem, K. J., and Pasternak, C. A.,** A novel method for measuring intracellular pH and potassium concentration, *Biosci. Rep.,* 3, 631, 1983.
23. **Enoki, Y., Kitayama, S., Ochiai, T., and Ikawa, Y.,** Red cell intracellular pH and the Bohr effect in human, swine and bovine blood, in *Proc. Intl. Semin. Blood Gas, Respiration and Pulmonary Vascular Function,* Chiba, 1979, 6.

Chapter 4A

ASSAYS OF INTRACELLULAR pH USING CHEMICAL PROBES: PRINCIPLES OF THE pH INDICATOR RESPONSE

Jan Slavík

TABLE OF CONTENTS

I. INTRODUCTION

There are several conceptually different groups of methods within this class of assays:

1. Distribution of a probe according to the difference between the extra- and intracellular pH values
2. Behavior of a probe introduced from the outside and "trapped" inside the cell
3. Behavior of an "intrinsic" probe

As each of these approaches will be treated here in separate chapters, (4B through 4E) it was thought to be useful to include an introductory chapter briefly summarizing the common properties of these pH-indicating substances. The specific problems of each technique will be left for the specialized chapters.

II. PRINCIPLES OF THE pH INDICATOR RESPONSE

The basic principles on which the various pH indicator methods are built are very similar. In fact, the methods differ only in the specific way in which the concentrations of the protonated and nonprotonated forms of the indicator are monitored.

All the substances used for intracellular pH measurements are weak electrolytes, and a description of their behavior was presented in the first chapter, being summarized for our purposes by the Henderson-Hasselbalch equation: $\text{pH} = \text{p}K_A + \log(c_{A^-}/c_{HA})$ for an acid or $\text{pH} = \text{p}K_B + \log(c_B/c_{BH^+})$ for a base (cf. Equations 44 and 46).

A. Intrinsic, Extrinsic, and Distributive pH Indicators

The suitable pH-indicating substance may already occur naturally inside the cell ("intrinsic" indicator), or it may be necessary to supply the cell with the indicator from outside ("extrinsic" indicator). The former case is common for a majority of NMR experiments (Chapter 4E), though in many NMR experiments extrinsic indicators also are introduced. Absorption (Chapter 4C) and fluorescence (Chapter 4D) methods rely almost exclusively on extrinsic indicators. The distribution techniques discussed in Chapter 4B use only distributive membrane-permeable indicators.

The indicators may be characterized as follows.

Intrinsic indicators — which occur naturally inside cells, may be handicapped by their inconvenient pH-indicating properties (e.g., unsuitable p*K* value, complicated detection because of a weak signal, etc.), low concentration of the indicator, or its uneven intracellular distribution resulting in discrimination of some regions and preference of others according to the local indicator concentration.

Extrinsic indicators — introduced into cells from the outside, may sometimes alter, by their very presence in the cell, the intracellular pH value and in some cases also influence metabolism of the cell.

Distributive indicators — have the same advantages and disadvantages as extrinsic pH indicators. While both forms of the extrinsic pH indicator are supposed to stay trapped inside the cell, the ideal distributive indicators should have the undissociated form permeant and the dissociated form impermeant (i.e., trapped inside).

B. Stability of Apparent p*K*

Both distribution and spectroscopic techniques rely in a similar way on the validity of Equation 44. Any change in intracellular p*K* produces an identical change in the derived pH_{in} value. From this point of view, the stability of p*K* is the alpha and omega of intracellular pH measurements.

Unfortunately, most of the factors influencing pK are unknown, and, thus, theoretical considerations are mostly based only on the Debye-Hückel theory and its extensions (cf. Chapter 1.V.). They describe effects of electrostatic interactions between the pH indicator and the medium. The interaction between the charged form of the indicator (one or both forms are charged, depending on the type of indicator) and the ions of the medium results in a change in the activity coefficient. Because the two forms differ in the electrical charge, the magnitude of this change is different for each form. The shift in pK is therefore given by the difference of logarithms of activity coefficients:

$$pK = \log \gamma_{acidic} - \log \gamma_{basic} \tag{77}$$

The Debye-Hückel theory and its extensions derive several more or less empirical approximations for activity coefficients. However, the most general type described by Equation 53 will be used here. This is the so-called Robinson-Stokes equation in which z will be the number of positive charges of the acid (and, hence, $z - 1$ on the dissociated, anionic, form).

It is certainly of help to assume that both the dissociated and undissociated forms of the indicator have the same coefficients, A, B, C, so that

$$\begin{aligned} pK &= A|z^2|\sqrt{I}/(1 + Ba\sqrt{I}) + CI - A|(z - 1)^2|\sqrt{I}/(1 + Ba\sqrt{I}) - CI \\ &= A(2z - 1)\sqrt{I}/(1 + Ba\sqrt{I}) \end{aligned} \tag{78}$$

This general equation classifies pH indicators into distinct categories with respect to the charges of their acidic and basic forms. Thus,

monoanionic ⇌ neutral

(e.g., $H_2PO_4^- + H^+ \rightleftharpoons H_3PO_4$

or $DMO^- + H^+ \rightleftharpoons DMO$)

$$\text{whence} \quad \Delta pK = A\sqrt{I}/(1 + Ba\sqrt{I}) \tag{79A}$$

dianionic ⇌ monoanionic

(e.g., $HPO_4^{2-} + H^+ \rightleftharpoons H_2PO_4^-$

or fluorescein^{2-} + H^+ ⇌ fluorescein$^-$)

$$\text{whence} \quad \Delta pK = 3A\sqrt{I}/(1 + Ba\sqrt{I}) \tag{79B}$$

trianionic ⇌ dianionic

(e.g., $PO_4^{3-} + H^+ \rightleftharpoons HPO_4^{2+}$

or carboxyfluorescein^{3-} + H^+ ⇌ carboxyfluorescein^{2-})

$$\text{whence} \quad \Delta pK = 5A\sqrt{I}/(1 + Ba\sqrt{I}) \tag{79C}$$

Therefore, if the apparent pK shifts were of purely electrostatic nature, the pH indicators should have pK shifts determined solely by their charge. For instance, indicators based on the transition between the neutral and charged forms should have a pK shift three times

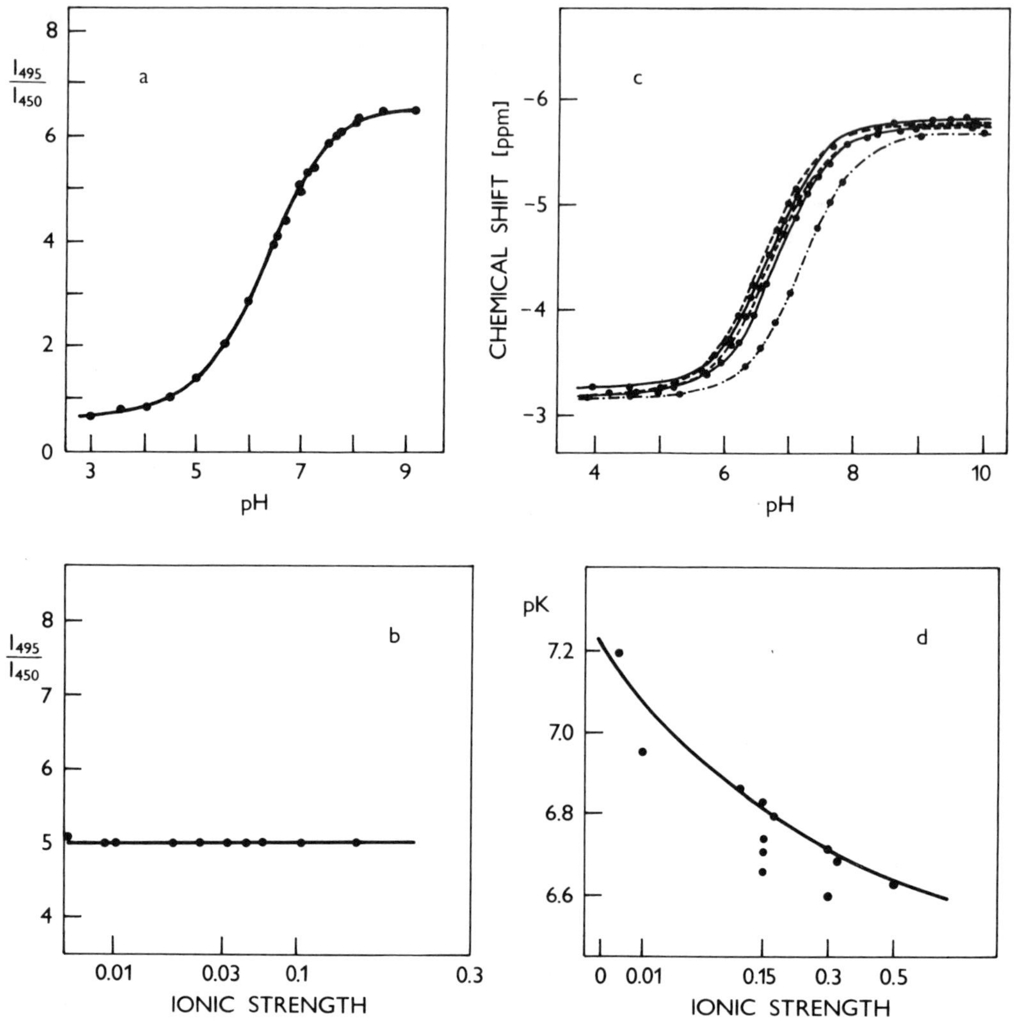

FIGURE 11. Dependence of the measured fluorescence-intensity ratios of fluorescence-isothiocyanate-labeled albumin (a,b) and chemical shift of orthophosphate (c.d) on pH and ionic strength of the medium. (Compiled from References 4 and 22.)

smaller than those based on the monoanionic-dianionic (or monocationic-dicationic) transitions. Furthermore, all pH indicators based on one type of transition should have exactly the same p*K* shift. Summarizing, any pH indicator should have a Δp*K* either $0.5\sqrt{I}/(1 + 1.5\sqrt{I})$ or $1.5\sqrt{I}/(1 + 1.5\sqrt{I})$ or $2.5\sqrt{I}/(1 + 1.5\sqrt{I})$. The p*K* shifts of weak base indicators should be of opposite sign but have the same value.

However, in practice, a number of deviations from this simple rule have been observed. Comparison of phosphate and fluorescein clearly shows striking differences, although both indicators fit Case 79b. As seen in Figure 11, the standard pH curves of phosphate are indeed sensitive to ionic strength whereas that of fluorescein is not (or apparently to a much lesser extent). The p*K* values derived from these calibration curves seem to follow a presumably parabolic shape in the case of phosphate, but are surprisingly linear with fluorescein.

More surprisingly, the pH-dependent chemical shift of methyl phosphonate shows little if any effect of ionic strength in contrast with phosphate indicators.[8,9]

A possible explanation for this may be the following. The approximations upon which

Equations 53 and 78 are based are still rather crude. For instance, they do not consider the exact distribution of the electrical charge within the molecule (which may be more localized in the case of the phosphate but largely delocalized in the case of the larger molecule of fluorescein). In the case of the fluorescein, there may be more than two forms of the indicator involved (cf. the rather complicated dissociation scheme of fluorescein in Chapter 4E, Figure 29), and consider the complexity if tautomeric forms according to Hantzsch's theory were introduced.

It should be concluded that the electrostatically induced shift of the apparent p*K* is only one among many possible shifts induced by the medium, or one of many aspects influencing the pH-calibration curve of the indicator. These effects may in some cases compensate each other (probably by coincidence) and result in an apparent independence of p*K* of the medium composition.

Be it as it may, derivation of the proper calibration curve characterizing the behavior of the pH indicator inside the cell is the cornerstone of any intracellular pH measurement and should be carefully performed. The different ways of calibration are discussed in detail at the end of this chapter.

Each measuring technique also has its specific effects which are discussed in the corresponding chapters. For instance, in fluorescence techniques, the shift in p*K* due to the excitation of the molecule may cause a perturbation of the already established equilibrium distribution between acidic and basic forms of the dye in the ground state. Because of the short duration of the excited state, the equilibration with respect to p*K** can never be completed, and the fragment of this process appears as a viscosity-dependent shift of fluorescence spectrum.

Likewise, in NMR, the chemical shift seems to comprise not only p*K*, but also some more complex effects (cf. Figures 73 and 75). Naturally, the formation of complexes with divalent ions veils further NMR measurement interpretations.

C. Tautomeric Forms

The spectroscopic techniques (absorption, fluorescence, and NMR spectroscopy) are based on a selective monitoring of intracellular concentrations of dissociated and undissociated forms of the pH-indicating substance, distinguishing between them by their absorption, fluorescence, and NMR spectra. In most cases, the techniques directly yield the ratio of concentrations of the two forms

$$R = c_{A^-}/c_{HA} \tag{80}$$

so that Equation 44 or 46 can be rewritten as

$$\mathrm{pH} = \mathrm{p}K_A + \log R \tag{81}$$

which represents the starting equation for all spectroscopic techniques.

The above description is usually applied to the interpretation of all data about intracellular pH measurement without any hesitation. Nevertheless, as it is based on Ostwald's theory of weak electrolyte indicators,[1] it should be (at least in the case of optical absorption measurements) complemented with the Hantzsch theory of color transitions.[2]

This theory, discussed in the chapter on absorption techniques, is based on the postulated existence of tautomeric forms of the indicator, which are responsible for the observed color change. Thus, this theory considers the equilibria between at least three different forms of the indicator and results in more complicated equations. However, the pH value is given again by R and a set of equilibrium constants (K_1, K_2, K_T)

$$pH = \log \frac{K_2 - K_1K_TR}{K_1K_2K_TR - K_1K_2} \tag{82}$$

where R is defined, as in Equation 80, as the ratio of concentrations of two differently colored species.

Nevertheless, in practice, empirically derived pH-standard curves are interpreted as being based on Ostwald's theory and on the use of various modifications of Equation 44 or 46. Usually, pK is replaced by an effective pK value calculated to best fit the experimental points of the standard curve. Some additional terms may appear as follows.

In single-wavelength measurement of absorption or fluorescence intensity or in measurement of NMR chemical shift, Equation 44 may be further transformed to

$$pH = pK_{eff} + \log \frac{I - I_a}{I_b - I} \tag{83}$$

where I denotes the measured value of absorbance, fluorescence intensity, or chemical shift; and I_a and I_b are the limiting values for pH $\ll$ pK and pH $\gg$ pK, respectively.[3-5]

In double-wavelength measurement of absorbance or fluorescence intensity, another equation is used:

$$pH = pK_{eff} + \log \frac{R - R_{min}}{R_{max} - R} - \log \frac{F_a}{F_b} \tag{84}$$

where R is the ratio measured, and R_{min} and R_{max} are the limiting values of the ratio at the extremes of acid and alkaline pH, respectively.[6,7]

With the exception of Equation 83 in the case of NMR chemical shifts both Equations 83 and 84 appear somewhat strange if confronted rigorously with Ostwald's theory. On the other hand, if the resulting equation of Hantzsch's theory (Equation 82) is rewritten to formally resemble these "empirically improved" equations, a striking similarity appears. Equation 82 can be expressed as

$$pH = \log K_2 + \log \frac{R - K_2/K_1K_T}{1/K_T - R} \tag{85}$$

and the empirically derived Equation 84 can be rewritten as

$$pH = \log(K_{eff}F_b/F_a) + \log \frac{R - R_{min}}{R_{max} - R} \tag{86}$$

This may lead to the conclusion that the improved versions of standard curves are, in fact, attempts to modify equations based originally on Ostwald's theory in the sense of Hantzsch's theory.

D. Location of the pH Indicator in the Cell

With distribution techniques, the fate of the pH-indicating molecule in a cell is rarely investigated and often remains unclear. The striking differences in the "accessible volume" alone, as stated by various authors, are puzzling.

In spectroscopic methods, the derived pH value reflects the weighted average of local pH values and local concentrations of the pH-indicating substance. Places with a higher concentration of the pH indicator are emphasized while places where no indicator is present do not contribute to the measured pH value.

An uneven intracellular distribution of the pH indicator may have several causes affecting both spectroscopic and distribution techniques.

First, the undissociated form of a monovalent pH indicator generally penetrates all membranes in a cell and tends to establish local concentration equilibria according to local transmembrane ΔpH. As a consequence, the indicator may accumulate within various subcellular compartments at significantly higher concentrations. The same principle applies to the charged forms of pH indicators, though to a lesser extent, as they are much less permeant. For instance, fluorescein is assumed to accumulate inside mitochondria, whereas carboxyfluorescein does not. Consequently, changes in intramitochondrial pH are reflected by fluorescein and do not appear in the signal of carboxyfluorescein.[30]

Second, charged forms of pH indicators may prefer some sites for purely electrostatic reasons. They are attracted by oppositely charged surfaces and often form aggregates with oppositely charged ions (ATP^{2-} and Mg^{2+}).

Third, the lipophilicity of the indicator may emphasize the contribution of the membrane volume itself by a bias in favor of the perimembrane pH value and in this way affect the measurement. Taking advantage of this effect, lipophilic derivatives of pH-sensitive fluorescent indicators are used.

However, in most cases a freely moving pH indicator evenly distributed inside the cell (in all compartments, if desired, or over the region of interest) is preferred. It should not interact with components constituting the cell nor interfere with any cellular process. These properties depend significantly on the concentration of the indicator employed. The choice of optimal intracellular concentration is one of the keys to success.

The recommended concentrations are in the range of 10 μ*M* to 1 m*M*. If the concentration is too low, a large proportion of indicator molecules may be bound. If the concentration is too high, the presence of the indicator may shift the intracellular pH value by its very presence or by perturbation of cell metabolism.

The extent of binding may be estimated by comparing the signal from the cell interior with that of the free indicator in solution. The maxima of absorption and fluorescence peaks of bound indicator are red-shifted (e.g., if 50% of fluorescein is bound to bovine serum albumin, the maximum of absorption shifts from 492 to 512 nm).[10] The polarization of fluorescence of freely moving molecules is almost zero, while in the case of entirely fixed molecules, it approaches values of 0.4. Also, the NMR signal gives broader lines for bound substances.

A simple way applicable to single cells and some other relatively simple objects is to determine the amount of intracellular probe at different osmotic values of the medium. If the amount of the probe changes hand in hand with the change of cell water volume, it may be assumed to be free in solution; if it shows a dependence on water volume containing a constant factor that can be extrapolated, binding can be assumed to occur.

In most cases the portion of bound indicator is negligible and does not affect the NMR, fluorescence, absorption, or distribution measurements. However, it should be stressed that the binding itself may change the pH-indicating properties, although this is an uncommon and certainly avoidable case.

E. Ways of Loading Cells with the Extrinsic pH Indicator

A reliable and convenient pH-indicating substance is rarely present inside the cell at a sufficiently high concentration. Hence, with the exception of most NMR techniques, an extrinsic membrane-impermeant pH indicator must be employed.

The main problem is how to bring the often membrane-impermeant compound across the membrane without damaging it or affecting the system (cell) measured. Various elaborate ways have been developed to solve the problem, reviewed, e.g., by Waddell and Bates[11] and Chambers and Chambers.[12]

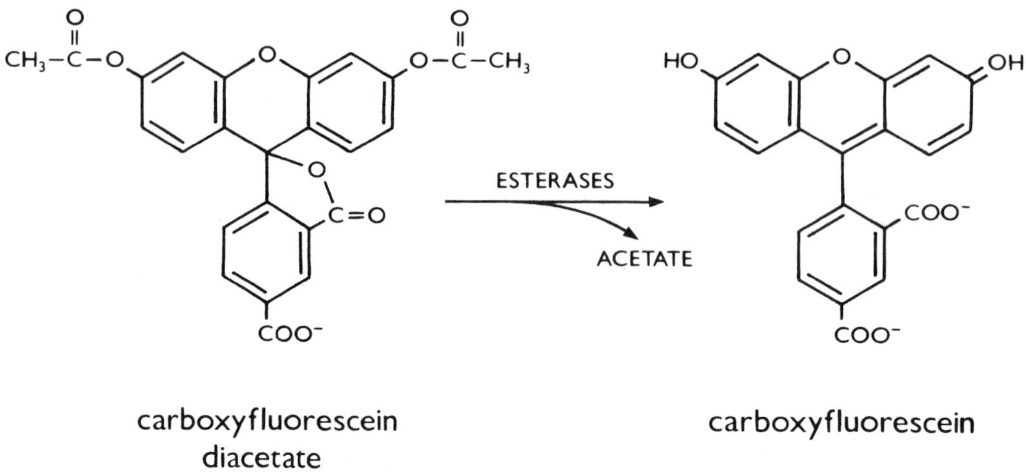

FIGURE 12. Hydrolysis of carboxyfluorescein diacetate.

The simplest way is represented by an incubation of cells in a solution of the pH-indicating substance. None of the indicators are in fact completely impermeant, so that after a certain time period a certain intracellular concentration of the pH-indicating substance is reached. The slow equilibration process may be speeded up by varying the external pH, based on the assumption that uncharged forms penetrate more easily and that the proportion of uncharged form is increased by pH (used, e.g., with carboxyfluorescein diacetate). However, those pH-indicating substances which easily penetrate inside the cell also show a high leakage rate out of the cell.

In some cells, endocytosis can be used.[9,13-16] Along the same line, fusion of cells with liposomes loaded with the pH indicator can solve the problem.[17-19] The transiently increased permeabilization of the membrane by an applied external voltage pulse is another attractive solution.[20,21]

A somewhat crude but widely used technique is the microinjection or pressure microinjection of the indicating substance into the cell.[13,15,16] However, this approach is feasible with larger cells only.[22-26]

Recently, a membrane-permeant precursor which is turned inside the cell into a pH-indicating substance has become extremely popular. This approach is now widely used in fluorescence and absorption measurements where a nearly nonfluorescent and colorless precursor is transformed "*in situ*" (intracellularly) into a highly fluorescent and absorbing substance which can later be employed as a pH indicator. The final substance is nearly membrane-impermeant and, thus, remains trapped inside the cell. (The approach is sometimes called "the technique of trapped indicator".) The precursor is, in almost all cases, an acylated derivative of an anionic indicator. The acyl group is cleaved off inside the cell by an unspecific enzymic reaction carried out mainly by esterases (Figure 12). The intracellular activity varies from cell to cell, but most cells exhibit a sufficient enzyme activity. Also, some of the acylated compounds decompose spontaneously at alkaline pH values.

Originally, the intracellular retention of fluorescein in cells incubated in a fluorescein-diacetate solution was introduced as a vital staining of cells to monitor the integrity and intactness of cell membranes. Later it was used in cell-sorting devices for separation of living and dead cells.[27,28] It was introduced quite recently into pH measurement.[29-31] Nowadays, a number of acetylated dyes suitable for intracellular pH measurement are offered, and many of them have already been used for intracellular pH measurement (e.g., fluorescein, carbofluorescein, methylumbelliferone).[32]

Similar ideas are applicable to NMR techniques, where pH-sensitive indicators are formed from intracellularly metabolized sugars; e.g., cells fed with fructose show a pH-sensitive fructose-1-phosphate peak.[33] Those fed with 2-deoxyglucose show a pH-sensitive peak of 2-deoxyglucose-6-phosphate.[34-36]

Recently, in ^{19}F NMR spectroscopy a fluorinated methyl ester of alanine was employed, intracellularly cleaved to free fluorinated amino acid, and used as a pH indicator.[37-41]

In ^{13}C NMR, where the natural abundance of the isotope is low, isotopically labeled substrates are employed to study the metabolism of the cell and could also be used for intracellular pH estimates.

F. Leak of pH Indicator from Cells

As was shown earlier in the text, it is difficult to bring the pH-indicating substance into the cell, but once inside it may be a problem to keep it there.

The pH indicator can leak out of the cell via numerous routes. The most important one is the diffusion of membrane-permeant forms of the indicator out of the cell. The indicator can also leave the cell using membrane carriers[42] (both passive and active transport). A microinjected indicator may leak out before the opening (injury) is properly sealed.

The possibility of a leak must always be considered, as none of the so-called membrane-impermeant pH indicators is totally impermeant. There is always at least one among its forms, especially the uncharged or less-charged form, that penetrates through the membrane. A typical example is the leak observed in the case of acetylated anionic indicators used for loading cells as described in the previous section. The cells typically show a twofold leakage: a slow leak of hydrolyzed probe and a fast leak of non-hydrolyzed probe. The former case is more significant in cells with poor esterase activity, the latter being in those with high esterase activity. The leakage rate in and out also depends on intracellular and extracellular pH (namely, via the already-mentioned changes in the proportion of charged and uncharged forms) and is influenced not only by intracellular enzymic activity, but also by pH-dependent stability of the acetylated compound (spontaneous decomposition) both inside and outside the cell.

Fortunately, there are several ways of minimizing the effects of leakage:

1. Slow down the rate of leakage.
2. Eliminate or strongly suppress the signal arising from the compound already leaked out.
3. Differentiate between intracellular and extracellular signals and separate them from each other.
4. Continuously remove the leaked compound.
5. Separate the cells immediately prior to measurement.

First, the rate of leakage depends on the type of cell and indicator. For instance, the half-life ranges from several weeks (carboxyfluorescein in liposomes)[43] to less than a minute (fluorescein in *Bacillus acidocaldarius*; see Figure 13). Generally, the use of indicators, both forms of which are charged, results in lower leakage. The decrease in temperature (storing cells on ice) may also decrease the rate of leakage.

Second, to decrease the intensity of the extracellular signal in NMR or EPR is relatively easy. The addition of paramagnetic ions (e.g., Mn[II] chloride or ferritin) results in complete disappearance of the extracellular signal, but the substance added must be impermeant.[45-47] In fluorescence techniques, addition of a potent extracellular fluorescence quencher represents a similar solution. However, in contrast to the above peak-broadening in magnetic resonance techniques which are generally applicable, the quenching of fluorescence is highly specific. For each dye another quencher may be needed. There is also a specific fluorescein fluorescence quenching rabbit antibody.

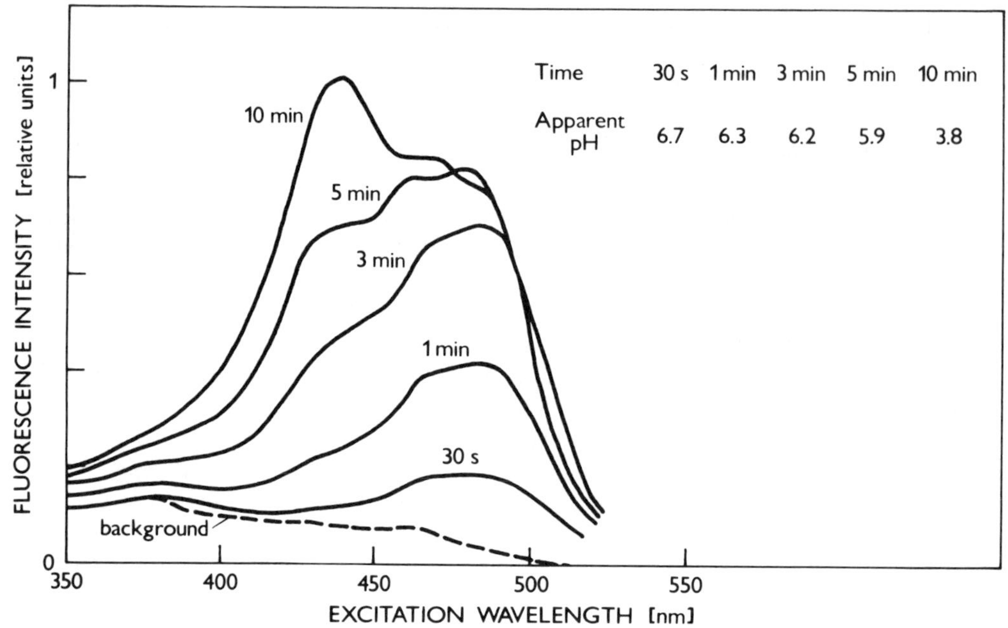

FIGURE 13. Effect of dye leakage on the estimated pH value in *Bacillus acidocaldarius*. Time indicates the duration of leakage. (Redrawn from Reference 30.)

Third, the microscopic observation in absorption or fluorescence allows spatial differentiation of the signal from the external and the internal sides of the cell. Moreover, as the extracellular volumes is relatively small, the extracellular signal is usually negligible. To some extent this applies also to flow cytometry experiments if the drops are small and the cells are large. In absorption measurement of thin (monolayer) tissue slices, the signal from the extracellular space may be blocked by a strongly absorbing dye not passing through the membrane. Therefore, no light from the extracellular space can appear and contribute to the measured intensity.[48]

Fourth, the concentration of the leaked indicator may be lowered by repeated washing or continual flow dialysis, continuous perfusion, rapid filtration or centrifugation, and possibly by enzymic degradation of the extracellular indicator.

Finally, the concentration of cells immediately prior to measurement improves the intracellular:extracellular signal ratio and helps to decrease the contribution of leaked indicator.

In cells with high leakage, combination of the above approaches can be useful. For instance, cells are stored on ice, then quickly recentrifuged immediately prior to the measurement. The pellet is daubed on a cuvette wall, covered with a microscopic cover slip, and immediately measured in frontal geometry fluorescence.

G. Derivation of the pH Standard Curve

The careful testing of pH-indicating substances cannot be overemphasized. Better knowledge of the indicator means more reliable results. Thus, each compound should be thoroughly tested in combination with the given cells, prior to pH measurement.

In distribution techniques, the stability of p*K* (cf. beginning of this chapter) and the fulfillment of postulated properties of permeating indicators (permeable-undissociated and impermeable-dissociated form no binding, no metabolization, are important and should be checked. A number of distribution techniques that work well in liposomes fail to give meaningful results in living cells (e.g., 9-aminoacridine). (See Chapter 4B, II.D.)

In spectroscopic techniques, the central point of calibration procedures is the determination of the standard pH curve. This procedure can be done *in vitro* in mineral buffers, in a cell extract, or *in situ*.

1. In a Buffer

The simplest case is to employ a series of buffers of different pH values. During the calibration procedure, the sample with the pH-indicating substance in a buffer is treated in the same way as the cells in the usual course of measurement. This procedure is often used as it is easy to perform and can be adapted to almost any kind of measurement (microcapillaries[7,48] or Petri dishes[31] filled with buffers in microscopy; capillaries or cuvettes in flow cytometry[49]). This approach using different buffer compositions allows the behavior of the indicator to be characterized and the nature of its response to effects other than pH to be elucidated.

2. In a Cell Extract

The use of cell-free extracts instead of artificially prepared (mostly mineral) buffers is preferable in cases of cells with an unusual cytoplasmic composition (e.g., *Halobacterium*). The extract is titrated with an acid or a base down and up, and, thus, the standard curve around the normal intracellular pH value is obtained.

3. In Situ

The *in situ* calibration uses several different approaches.

The microinjection of a strong buffer of known pH value is based on the assumption that the buffering capacity of the buffer is sufficiently stronger than that of the cell and that the final intracellular pH will adjust to that of the injected buffer.[22] However, the increase in ionic strength due to injection of a strong buffer may distort the results.

Instead of microinjection, ionophores can be applied to cells suspended in buffers to different pH values. The standard curve is constructed on the assumption that, after addition of the ionophore (nigericin, FCCP, or CCCP), the intracellular pH finally reaches the extracellular pH. Some doubt remains whether this assumption is always true.[22,29,30,50,51]

A fine modification uses suspensions of cells in a series of buffers of different pH values. The external pH is varied until no change in the pH-indicator signal appears upon eventual cell lysis.[51-53] However, as in the above-mentioned injection of buffers, the change of ionic strength upon lysis may distort the results; it is always an advantage if two independent techniques can be used.

In conclusion to the calibration section, it may again be emphasized that the properties of the indicator substance are often more important than the technique used for its monitoring. The possible deviations from the idealized, postulated behavior of the pH-indicating substance sometimes present more difficult problems than the measurement per se, and, thus, only well-tested indicators should be employed.

REFERENCES

1. **Ostwald, W.,** Grundsätzliches zur messenden Farbenlehre, *Akad. Wiss. (Berlin),* 1929 and 1937.
2. **Hantzsch, A.,** Influence of solvents on the absorption of light by dissolved substances, *Berichte,* 50, 1413, 1918.
3. **Jacobus, W. E., Pores, I. H., Lucas, S. K., Kallman, C. H., Weisfeldt, M. L., and Flaherty, J. T.,** The role of intracellular pH in the control of normal and ischemic myocardial contractility: a ^{31}P NMR and mass spectrometry study, in *Intracellular pH: Its Measurement, Regulation, and Utilization in Cellular Functions,* Nuccitelli, R. and Deamer, D. W., Eds., Alan R. Liss, New York, 1982, 537.

4. **Seo, Y., Murakami, M., Watari, H., Imai, Y., Yoshizaki, K., Nishikawa, H., and Morimoto, T.,** Intracellular pH determination by a ^{31}P-NMR technique: the second dissociation constant of phosphoric acid in a biological system, *J. Biochem.*, 94, 729, 1983.
5. **Geisow, M. J.,** Fluorescein conjugates as indicators of subcellular pH, *Exp. Cell Res.*, 150, 29, 1984.
6. **Bashford, C. L.,** Spectrophotometry and fluorimetry of cellular compartments, in *Spectrophotometry and Spectrofluorimetry — a Practical Approach,* Harris, D. A. and Bashford, C. L., Eds., IRL Press, Oxford, 1987, chap. 5.
7. **Graber, M. L., DiLillo, D. C., Bradford, F. L., and Pastoriza-Munoz, E.,** Characteristics of fluoroprobes for measuring intracellular pH, *Anal. Biochem.*, 156, 202, 1986.
8. **Weinstein, J. N., Ralston, E., Leserman, L. D., Klausner, R. D., Dragsten, P., Henkart, P., and Blumenthal, R.,** Self-quenching of carboxyfluorescein fluorescence: uses in studying liposome stability and liposome-cell interaction, in *Liposome Technology,* Gregoriadis, G., Ed., CRC Press, Boca Raton, FL, 1984, chap. 13.
9. **Ohkuma, S. and Poole, B.,** Fluorescence probe measurement of the intralysosomal pH in living cells and the perturbation of pH by various agents, *Proc. Natl. Acad. Sci. U.S.A.*, 75, 3327, 1978.
10. **Udkoff, R. and Norman, A.,** Polarization of fluorescein fluorescence in single cells, *J. Histochem. Cytochem.*, 27, 49, 1979.
11. **Waddell, W. J. and Bates, R. G.,** Intracellular pH, *Physiol. Rev.*, 49, 285, 1969.
12. **Chambers, R. and Chambers, E. L.,** Hydrogen-ion concentration of cell components, in *Explorations into the Nature of the Living Cell,* Harvard University Press, Cambridge, MA, 1961, 141.
13. **Geisow, M. J.,** Fluorescein conjugates as indicators of subcellular pH, *Exp. Cell. Res.*, 150, 29, 1984.
14. **Geisow, M. J. and Evans, W. H.,** pH in the endosome: measurements during pinocytosis and receptor-mediated endocytosis, *Exp. Cell Res.*, 150, 36, 1984.
15. **Heiple, J. M. and Taylor, D. L.,** Intracellular pH in single motile cells, *J. Cell. Biol.*, 86, 885, 1980.
16. **Tanasugarn, L., McNeil, P., Reynolds, G. T., and Taylor, D. L.,** Microspectrofluorometry by digital image processing: measurement of cytoplasmic pH, *J. Cell. Biol.*, 98, 717, 1984.
17. **Weinstein, J. N., Yoshikami, S., Henkart, P., Blumenthal, R., and Hagins, W. A.,** Transfer of a fluorescent dye from lipid vesicles to cells, *Biophys. J.*, 16, 104a, 1976.
18. **Straubinger, R. M., Hong, K., Friend, D. S., and Papahadjopoulos, D.,** Endocytosis of liposomes and intracellular fate of encapsulated molecules, *Cell,* 32, 1069, 1983.
19. **Kano, K. and Fendler, J. H.,** Pyranine as a sensitive pH probe for liposome interiors and surfaces. pH gradients across phospholipid vesicles, *Biochim. Biophys. Acta,* 509, 289, 1978.
20. **Gylfe, E., Berggren, P. O., and Rorsman, P.,** Measurements of cytoplasmic pH in insulin-releasing cells with intracellularly trapped phenol red, *Biomed. Res.*, 7, 139, 1986.
21. **Knight, D. E. and Baker, P. F.,** Calcium-dependence of catecholamine release from bovine adrenal medullary cells after exposure to intense electric fields, *J. Membr. Biol.*, 68, 107, 1982.
22. **Heiple, J. M. and Taylor, D. L.,** An optical technique for measurement of intracellular pH in single living cells, in *Intracellular pH: Its Measurement, Regulation and Utilization in Cellular Functions,* Nuccitelli, R. and Deamer, D. W., Eds., Alan, R. Liss, New York, 1982, 21.
23. **Chambers, R., Pollack, H., and Hiller, S.,** The protoplasmic pH of living cells, *Proc. Soc. Exp. Biol. Med.*, 24, 760, 1927.
24. **Lisman, J. E. and Strong, J. A.,** The initiation of excitation and light adaptation in Limulus ventral photoreceptors, *J. Gen. Physiol.*, 73, 219, 1979.
25. **Brown, J. E., Brown, P. K., and Pinto, L. H.,** Detection of light-induced changes of intracellular ionized calcium concentration in Limulus ventral photoreceptors using Arsenazo III, *J. Physiol.*, 267, 299, 1977.
26. **Baylor, S. M., Chandler, W. K., and Marshall, M. W.,** Optical measurements of intracellular pH and magnesium in frog skeletal muscle fibers, *J. Physiol.*, 331, 105, 1982.
27. **Rotman, D. and Pappenmaster, B. W.,** Membrane properties of living mammalian cells as studied by enzymatic hydrolysis of fluorogenic esters, *Proc. Natl. Acad. Sci. U.S.A.*, 55, 134, 1966.
28. **Bonner, W. A., Hulett, H. R., Sweet, R. G., and Herzenberg, L. A.,** Fluorescence activated cell sorting, *Rev. Sci. Instrum.*, 43, 404, 1972.
29. **Thomas, J. A., Buschbaum, R. N., Zimniak, A., and Racker, E.,** Intracellular pH measurements in Ehrlich ascites tumor cells utilizing spectroscopic probes generated in situ, *Biochemistry,* 18, 2210, 1979.
30. **Thomas, J. A., Kolbeck, P. C., and Langworthy, T. A.,** Spectroscopic determination of cytoplasmic and mitochondrial pH transitions using trapped pH indicators, in *Intracellular pH: Its Measurement, Regulation and Utilization in Cellular Functions,* Nuccitelli, R. and Deamer, D. W., Eds., Alan R. Liss, New York, 1982, 105.
31. **Slavík, J.,** Intracellular pH topography: determination by a fluorescent probe, *FEBS Lett.*, 156, 227, 1983.
32. **Thomas, J. A.,** Intracellularly trapped pH indicators, in *Optical Methods in Cell Physiology,* Deweer, P. and Salzberg, B. M., Eds., John Wiley & Sons, New York, 1986, 311.
33. **Iles, R. A., Griffiths, J. R., Stevens, A. N., Gadian, D. G., and Porteous, R.,** Effects of fructose on the energy metabolism and acid-base status of the perfused-starved rat liver, *Biochem. J.*, 192, 191, 1980.

34. **Navon, G., Ogawa, S., Shulman, R. G., and Yamane, T.,** ^{31}P NMR studies of Ehrilich ascites tumor cells, *Proc. Natl. Acad. Sci. U.S.A.*, 74, 87, 1977.
35. **Bailey, I. A., Williams, S. R., Radda, G. K., and Gadian, D. G.,** Activity of phosphorylase in total global ischaemia in the rat heart: a ^{31}P NMR study, *Biochem. J.*, 196, 171, 1981.
36. **Civan, M. M., Williams, S. R., Gadian, D. G., and Rozengurt, E.,** ^{31}P NMR analysis of intracellular pH of Swiss Mouse 3T3 cells, *J. Membr. Biol.*, 94, 55, 1986.
37. **Deutsch, C., Taylor, J. S., and Wilson, D. F.,** Regulation of intracellular pH by human peripheral blood lymphocytes as measured by ^{19}F NMR, *Proc. Natl. Acad. Sci. U.S.A.*, 79, 7944, 1982.
38. **Deutsch, C., Taylor, J. S., and Prize, M.,** pH homeostasis in human lymphocytes — modulation by ions and mitogen, *J. Cell Biol.*, 98, 885, 1984.
39. **Taylor, J. S. and Deutsch, C.,** Fluorinated alpha-methylamino acids as ^{19}F NMR indicators of intracellular pH, *Biophys. J.*, 43, 261, 1983.
40. **Civan, M. M., Lin, L. E., Peterson-Yantorno, K., Taylor, J., and Deutsch, C.,** Intracellular pH of perfused single frog skin: combined ^{19}F and ^{31}P NMR analysis, *Am. J. Physiol.*, 16, C506, 1984.
41. **Kashiwagura, T., Deutsch, C., Taylor, J., Erecińska, M., and Wilson, D. F.,** Dependence of gluconeogenesis, urea synthesis, and energy metabolism of hepatocytes on intracellular pH, *J. Biol. Chem.*, 259, 237, 1984.
42. **Prosperi, E., Croce, A. C., Bottiroli, G., and Supino, R.,** Flow cytometric analysis of membrane permeability properties influencing intracellular accumulation and efflux of fluorescein, *Cytometry*, 7, 70, 1986.
43. **Weinstein, J. N., Yoshikami, S., Henkart, P., Blumenthal, R., and Hagins, W. A.,** Lysosome-cell interaction: transfer and intracellular release of a trapped fluorescent marker, *Science*, 195, 489, 1977.
44. **Stewart, I. M., Chapman, B. E., Kirk, K., Kuchel, P. W., Lovric, V. A., and Raftos, J. E.,** Intracellular pH in stored erythrocytes: refinement and further characterization of the ^{31}P NMR methylphosphonate procedure, *Biochem. Biophys. Acta*, 885, 23, 1986.
45. **Mehlhorn, R. J., Candau, P., and Packer, L.,** Measurements of volumes and electrochemical gradients with spin probes in membrane vesicles, *Methods Enzymol.*, 88, 751, 1982.
46. **Mehlhorn, R. J. and Probst, I.,** Light-induced pH gradients measured with spin-labeled amine and carboxylic acid probes: application to Halobacterium halobium cell envelope vesicles, *Methods Enzymol.*, 88, 334, 1982.
47. **Herbst, M. and Piontek, P.,** Über den Verlauf des intracellularen pH-Wertes des Skeletmuskels während der Kontraktion, *Pflügers Arch.*, 335, 213, 1972.
48. **Sernetz, M. and Thaer, A.,** Microcapillary fluorometry and standardization for microscope fluorimetry, in *Fluorescence Techniques in Cell Biology*, Thaer, A. and Sernetz, M., Eds., Springer-Verlag, Berlin, 1973, 41.
49. **Gerson, D. F.,** Determination of intracellular pH changes in lymphocytes with 4-methylumbelliferone by flow microfluorimetry, in *Intracellular pH: Its Measurement, Regulation and Utilization n Cellular Functions*, Nuccitelli, R. and Deamer, D. W., Eds., Alan R. Liss, New York, 1982, 125.
50. **Kaplan, A. E. and Bunow, M. R.,** Spectrophotometric determination of intracellular pH with cultured rat liver epithelial cells, *J. Histochem. Cytochem.*, 34, 749, 1986.
51. **Paradiso, A. M., Negulescu, P. A., and Machen, T. E.,** Na^+-H^+ and Cl^--OH (HCO_3^-) exchange in gastric glands, *Am. J. Physiol.*, 25, G524, 1986.
52. **Babcock, D. F.,** Internal ion concentrations coupled to intracellular fluorescein chromophore with ionophores, *Fed. Proc.*, 40, 1785, 1981.
53. **Graber, M. L., Dixon, T. E., Coachman, D., Herring, A. R., Gardner, T., and Pastoriza-Munoz, E.,** Fluorescence identifies an alkaline cell in turtle urinary bladder, *Am. J. Physiol.*, 250, F159, 1986.

Chapter 4B

ASSAYS OF INTRACELLULAR pH USING CHEMICAL PROBES: DISTRIBUTION TECHNIQUES

Jan Slavík and Arnošt Kotyk

TABLE OF CONTENTS

I. THEORETICAL ASPECTS

A. Introduction

The most notable features of this widely used and traditional method are more than 60 years of experience and a simplicity of execution. The measurement of intracellular pH is based on monitoring the equilibrium distribution of a membrane-permeant weak electrolyte between the intracellular and extracellular spaces.

B. Basic Principle

The estimation of intracellular pH is based on the assumption that the uncharged form of a weak electrolyte is permeant while the charged form is impermeant. Consequently, weak acids are accumulated inside cells after addition to cell suspension if the intracellular pH is lower than the extracellular pH. In analogy, weak bases are accumulated if the intracellular pH is higher than the external one. After a certain time, which is needed for the establishment of the steady-state equilibrium concentration of the indicating electrolyte, the intracellular pH value is calculated from the ratio of analytical concentrations of the electrolyte in the external and internal fluids, the tabulated p*K* value, and the extracellular pH value. Usually, the analytical concentration of the pH indicator, i.e., $c_{HA} + c_{A-}$ with an acid, or $c_B + c_{BH+}$ with a base, is determined either extracellularly or intracellularly; and the counterpart (i.e., extracellular or intracellular concentration, respectively) is determined as the difference, using the total analytical amount of the indicator as the starting point.

The theory of distribution measurements is derived as follows.

The dissociation degree of a weak electrolyte is given by its valency, apparent p*K*, and ambient pH (cf. Chapter 1, Equations 44 and 46). For a monobasic weak acid in the intracellular and extracellular spaces, we obtain

$$pH_{in} = pK_{in} + \log(c_{A^-_{in}}/c_{HA_{in}}) \quad (87a)$$

$$pH_{out} = pK_{out} + \log(c_{A^-_{out}}/c_{HA_{out}}) \quad (87b)$$

The basic postulate of the distribution techniques (permeability of the nondissociated and impermeability of the dissociated form) may be expressed as

$$c_{HA_{in}} = c_{HA_{out}} \quad (87c)$$

For the sake of simplicity, it may be assumed that the apparent dissociation constant does not significantly differ between intracellular and extracellular spaces (cf. Chapter 4A) so that

$$pK_{out} = pK_{in} \quad (87d)$$

If conditions in Equations 87c and 87d are fulfilled, the value of intracellular pH may be calculated.

This calculation is then based on the following considerations. What can be determined is the ratio of analytical concentrations, defined for an acid by

$$R_A = (c_{HA_{in}} + c_{A^-_{in}})/(c_{HA_{out}} + c_{A^-_{out}}) = c^T_{A_{in}}/c^T_{A_{out}} \quad (88a)$$

for a base, then, by

$$R_B = (c_{B_{in}} + c_{BH^+_{in}})/(c_{B_{out}} + c_{BH^+_{out}}) = c^T_{B_{in}}/c^T_{B_{out}} \tag{88b}$$

The values of c_{A^-} are computed from the dissociation constant of HA so that

$$c_{A^-_{out}} = c_{HA_{out}}K_A/c_{H^+_{out}} \quad \text{and} \quad c_{A^-_{in}} = c_{HA_{in}}K_A/c_{H^+_{in}} \tag{89}$$

Using Equation 88a and the conditions that $c_{HA_{in}} = c_{HA_{out}}$ and $K_{A_{in}} = K_{A_{out}}$, we have $c_{H^+_{in}} = c_{H^+_{out}}c_{A^-_{out}}/c_{A^-_{in}}$ as well as $c^T_{A_{out}} = c_{HA}\,(1 + K_A/c_{H^+_{out}})$ and $c^T_{A_{in}} = c_{HA}(1 + K_A/c_{H^+_{in}})$. From this,

$$R_A = (1 + K_A/c_{H^+_{in}})/(1 + K_A/c_{H^+_{out}}) \tag{90}$$

Combining these expressions to eliminate c_{A^-} and making use of the relation $x = 10^{\log x} = 10^{-px}$ leads to the final formula for intracellular pH based on the distribution of a weak acid[1]

$$pH_{in} = pH_{out} + \log[R_A(1 + 10^{pK_A - pH_{out}}) - 10^{pK_A - pH_{out}}] \tag{91a}$$

or[2]

$$pH_{in} = \log[R_A(10^{pK_A} + 10^{pH_{out}}) - 10^{pK_A}] \tag{91b}$$

By analogy, proceeding from the dissociation constant of a weak monobasic base ($BH^+ = B + H^+$), one can derive the formula for intracellular pH based on the distribution of a weak base[3]

$$pH_{in} = pH_{out} - \log[R_B(1 + 10^{pH_{out} - pK_{BH}}) - 10^{pH_{out} - pK_{BH}}] \tag{92a}$$

(here $pK_{BH} = pK_W - pK_B$; see Equation 26).

or[2]

$$pH_{in} = -\log[R_B(10^{-pK_B} + 10^{-pH_{out}} - 10^{-pK_B}] \tag{92b}$$

Equations 91a,b and 92a,b may be simplified with little error introduced if the pK value of the indicator is appreciably different from pH_{out}. Thus, even if the difference is of only one pH unit (say, $pK = 4$ and $pH_{out} = 5$), the pH_{in} resulting from the simplified formula would differ from the true one by less than 0.05 pH unit. Then,

$$pH_{in} = pH_{out} + \log(c^T_{A_{in}}/c^T_{A_{out}}) = pH_{out} - \log(c^T_{B_{in}}/c^T_{B_{out}}) \tag{93a}$$

or

$$\Delta pH = \log(c^T_{A_{in}}/c^T_{A_{out}}) = \log(c^T_{B_{out}}/c^T_{B_{in}}) \tag{93b}$$

The application of dibasic weak electrolytes such as fluorescent diamines requires a modification of the above given equations (cf. Equations 96a,b).

C. The Effect of Cell Compartmentation

The distributive indicator is presumed to permeate not only through the cell membrane, but also through all other membranes within the cell. As a result, it is distributed in all intracellular compartments at concentrations corresponding to the particular pH values of these compartments. Mathematical analysis shows that the intracellular distribution of a weak acid is principally different from that of a weak base, and also that the distribution of a monovalent weak electrolyte differs from that of a divalent one.

The measured "average" pH value is, thus,[4,5]

$$\overline{pH}_{acid} = \log \sum_{j=1}^{n} f_j \, 10^{pH_j} \tag{94a}$$

when indicated by a weak monovalent acid and

$$\overline{pH}_{base} = -\log \sum_{j=1}^{n} f_j \, 10^{-pH_j} \tag{94b}$$

when indicated by a weak monovalent base (f_j denotes the fractional volume of the j-th compartment, and pH_j the internal pH of that compartment). Obviously, in compartmental systems, weak acids give a higher pH reading than weak bases, thus bracketing the volume-averaged internal pH value

$$\overline{pH}_{in} = \sum_{j=1}^{n} f_j \, pH_j \tag{94c}$$

in the following way:

$$\overline{pH}_{acid} > \overline{pH}_{in} > \overline{pH}_{base} \tag{95}$$

There are two important practical consequences of this phenomenon. First, a simultaneous application of a weak monovalent acid and a weak monovalent base makes it possible to attain more accurate estimates of pH_{in}.[6] In this connection, it should be emphasized that the pH_{in} value defined by Equation 94c is different from that obtained from a cell homogenate, as the latter comprises not only the volumes and pH values of the compartments, but also their buffering powers.

Second, in a system with two compartments, the application of both a weak monovalent acid and a weak monovalent base or a weak monovalent base together with a divalent one makes it possible to estimate the pH values in both compartments at the same time. For instance, in this way the values of cytoplasmic and intrathylakoidal pH in chloroplasts and cyanobacteria were estimated.[7-10] The separate monitoring of the two indicators can be done either consecutively or simultaneously, using differences in fluorescence wavelengths or different radioactive labeling.

II. DISTRIBUTIVE INDICATORS OF pH

The intracellular and extracellular concentrations of pH-indicating weak electrolytes can be monitored in various ways. The indicator can be labeled with a radioactive, fluorescent, or EPR tag or its NMR properties can be used, or the indicator can be analyzed chemically. The usual way is to use radioactively labeled indicators (most often ^{14}C-labeled) though a number of results have been obtained from colored or fluorescent substances. Recently, spin-labeled indicators and ^{31}P NMR spectroscopy were successfully introduced. (It should be pointed out, however, that the applications of optical and magnetic resonance techniques discussed here are limited only to the determination of intracellular and extracellular concentration of a distributive indicator. Application of these techniques to monitoring pH-dependent spectral changes of an intracellularly trapped indicator is described in detail in Chapters 4C through 4E.)

When considering the choice of a suitable indicator and the particular technique for monitoring its concentration, it appears that the indicator choice is more important. The most important criterion is how an indicator approaches the ideal behavior of a distributive indicator (no binding, no metabolism, no permeability of the charged form, etc.).

Regarding the choice of monitoring techniques, the use of radiolabeled compounds seems to be the most straightforward, while other techniques, however elegant and simple they may appear, can introduce additional difficulties (such as unclear nature of fluorescence quenching, instability of spin radicals, etc.).

A. Radioactively Labeled Indicators

The number of weak acids and bases that have been successfully employed as distributive indicators of intracellular pH is quite high. In most cases the indicators were labeled with ^{14}C, and centrifugation or filtration (more recently flow dialysis[11,12]) was used for separation of cells from the suspending medium. In some cases where spatial differentiation is required, autoradiographic techniques may be employed.

1. Weak Acids

The overwhelming majority of today's papers deal with ^{14}C-labeled 5,5-dimethyl-2,4-oxazolidinedione[13-22] (DMO; pK ~6.1 to 6.3) which apparently meets almost all requirements on an ideal distributive indicator. Other frequently used indicators are lactic acid[23-26] (pK ~3.7), acetic acid[11,27] (pK ~4.8), propionic acid[28-30,44] (pK ~4.9), salicylic acid[31] (pK ~3.0), 2,4-dinitrophenol[32] (pK ~4.9), dimethadione[33] (2,4-oxazolidinedione), and phenobarbital[34] (pK ~7.8). One of the most popular weak acids of the past was bicarbonate[35,36] ($CO_2 \rightleftharpoons HCO_3^-$ with pK ~6.3), later abandoned as too complicated to work with in comparison to the other indicators.[37]

2. Weak Bases

Nicotine[6,38] (pK ~8.2) appears to be the best choice among weak bases though in comparison to DMO it is far from ideal (cf. Section 3c). Other weak bases include lidocaine[39] (pK ~7.9), morpholine[40] (pK ~8.5), procaine[34,39] (pK ~9.0), morphine[41] (pK ~9.9), methylamine[14,22,42-45] (pK ~10.7), trimethylamine[41] (pK ~9.7), and ethylamine[46] (pK ~10.5).

B. Indicators That Can Be Monitored With Ion-Selective Electrodes

Ammonium ($NH_3 \rightleftharpoons NH_4^+$; p$K$ ~9.2) was the first weak base employed for intracellular pH measurement.[47] The classical approach used was to follow ammonium concentration analytically,[48-51] but lately an elegant way of monitoring the extracellular concentration with an ion-selective ammonium electrode was demonstrated.[52,53] It turned out, however, that the low selectivity and sensitivity of the electrode may limit the accuracy of the intracellular pH measurement altogether.[52]

C. Colored Indicators

Concentrations of a colored weak electrolyte can be easily and continuously followed by optical absorption measurement. Generally, the intracellular amount is best obtained from the difference between the value in a solution without cells and the value in the supernatant or filtrate after cells have been added. When working with a pH-sensitive dye which also exhibits a pH-dependent color transition, care must be taken that the supernatant or filtrate absorbance be read only after bringing the pH well above or below the pK with a drop of hydroxide or acid solution.

A number of results have been obtained with colored substances, both in unicellular organisms[1] or organelles.[54] An example here is the application of bromophenol blue.[1,55,56]

These approaches use the dye as a distributive weak electrolyte and should not be confused with experiments in which the pH-dependent color transition is employed as a means of determination of intracellular pH by a color change of an intracellularly trapped pH indicator. Intracellular pH indicators are described in Chapter 4C.

D. Fluorescence Quenching

The use of fluorescence for monitoring intracellular concentrations of a distributive fluorescent indicator is very popular, as it gives a continuous signal and is relatively simple to perform. The fluorescence-quenching technique assumes that the fluorescence of the internally located dye is totally quenched and that the residual fluorescence arises from the external dye only.[57] Therefore, the intensity of fluorescence before and after the uptake of the dye gives the amount of internalized dye, and, thus, intracellular pH may be calculated. Because so far only weak bases with pK's much greater than pH_{in} have been used, the simplified type of basic equation (cf. Equation 93b) may be applied. The pH difference across the membrane is then related to the fraction Q of the total initial fluorescence quenched during the equilibration by the equation

$$\Delta pH = \log(1 - Q)/Q + \log(w/V) \qquad (96a)$$

if a monobasic weak base is used, or

$$\Delta pH = 0.5 \log(1 - Q)/Q + \log(w/V) \qquad (96b)$$

for a dibasic weak base. In both equations, w/V denotes the ratio of intracellular water volume to the total extracellular volume of the sample.

In contrast to the fine-looking theory, reality is not so satisfying. The mechanism of fluorescence quenching of the internalized probe is unclear. It is assumed that some kind of concentration-dependent self-quenching by an excitation energy transfer takes place.[58] Unfortunately, the quenching effect starts at relatively high concentrations of the internalized dye[57] which may already alter the measured intracellular pH. More important, detailed studies of experimental results show that none of the indicators used so far meets the requirements placed on a distributive indicator if applied in living cells. The only case when the technique can be reliably used is in liposomes. The use of fluorescence quenching for intracellular pH measurement in real cells, however popular it may seem to be, cannot be recommended.[59-63] All the results should be considered as qualitative or semiquantitative only. The following dyes are in use.

9-Aminoacridine (9AA) — This monoamine represents a monobasic base with pK of about 10.0. Optimum excitation is at about 400 nm. The emission spectrum consists of two peaks (430 and 460 nm).[22,53,57-60,62,64-71]

9-(4-Diethylamino-1-methylbutylamino)-6-chloro-2-methoxyacridine — This diamine dye, known as atebrin or quinacrine, behaves as a dibasic weak base with pK_1 ~7.9 and pK_2 ~10.5. The excitation seems to be optimum between 420 and 450 nm. Fluorescence maximum appears at about 500 nm.[59,67,70,72]

3,6-*bis*(Dimethylamino)acridine — The concentration-dependent quenching of this dye, commonly known as acridine orange, may also be used for direct intracellular pH measurement. With rising concentrations, the monomeric fluorescence (maximum around 530 to 535 nm) is quenched while the polymeric fluorescence (maximum 640 nm) rises. Maximum absorption lies at 489 nm.[45,70,73]

9-Amino-6-chloro-2-methoxyacridine (9ACMA) — This monoamine has a pK of 9.0 and an excitation optimum at 420 nm.[59]

E. Spin-Labeled Indicators

In analogy to radioactive labeling, spin labels and EPR spectroscopy can be successfully employed for monitoring the distribution of substances between the cell and the external medium. The technique evaluates the EPR signal of a spin-labeled substance before and after addition of the "quencher" which does not permeate the cell membrane and is capable of quenching the entire signal of the extracellularly located substance. In comparison to radioactively labeled electrolytes, distribution data are available almost immediately, including data on the cell water volume. The sensitivity of EPR spectroscopy allows spin-labeled acids or bases to be employed even at the "wrong" side of the p*K* value.[74] As this technique is still under development, no definite conclusions can be made on its basis.

The most successful indicators tested so far are nitroxide derivatives of carboxylic acid and amines. They have been tested in membrane vesicles[75-77] and also in intact cells.[74,78] The main drawback *in vivo* is the signal loss of nitroxide spin probes in the reducing environment of the cell interior.[74]

The nitroxide-labeled weak electrolytes suitable for intracellular pH measurement are commercially available from Molecular Probes.[79] They include derivatives of carboxylic acid: 2,2,5,5-tetramethyl-3-pyrroline-1-oxyl-3-carboxylic acid (T 516)[76,77] and 2,2,5,5-tetramethylpyrrolidin-1-oxyl-3-carboxylic acid (T 517);[74] amines: 4-amino-2,2,6,6-tetramethylpiperidine-1-oxyl (A 502, TEMPAMINE)[76,77,80,83] and 3-amino-2,2,5,5-tetramethylpyrrolidin-1-oxyl (A 519, p*K* ~7.55);[74] neutral molecules for the determination of cell volume: 4-oxo-2,2,6,6-tetramethylpiperidine-1-oxyl (O 503, TEMPONE)[76,77,83] and 2,2,5,5-tetramethylpyrrolidin-1-oxyl (HO 877).[74]

Membrane-impermeant quenchers of the extracellular (or extravesicular) signal used were ferricyanide[75,76,83] and nickel chelates.[74]

F. NMR Spectroscopy

NMR spectroscopy is capable of distinguishing between the signals originating inside and outside the cell. Therefore, if a suitable permeant indicator is applied, estimates of intracellular and extracellular concentrations of this indicator and, thus, pH_{in} assays are possible as well as the determination of intracellular volume.[84,85] This approach is so clear and straightforward that it is surprising that, with the exception of methyl phosphonate,[86,87] no other such attempts to use distributive indicators of NMR have yet been reported.

Nevertheless, if a distributive indicator is added to cells, the NMR spectrum consists of two peaks: one arising from the intracellularly located substance, the other from the extracellular one. The size of the peaks (integral of the peak area) is proportional to the concentration of the intracellularly and extracellularly located substance, whereas the exact position of the peaks (chemical shift value) corresponds to the intracellular and extracellular pH values. Unfortunately, the chosen indicator, methyl phosphonate, does not fulfill at all the conditions for a distributive pH indicator because its charged forms are extremely membrane permeable, and the final distribution corresponds more to the membrane potential than to pH values.[86]

The dimethyl derivative of methylphosphonate is suitable for estimates of the average cell volume and can be followed by both ^{31}P NMR and ^{1}H NMR.[85]

III. EXPERIMENTAL ASPECTS

A. Determination of Extracellular and Intracellular Volume

The calculation of intracellular pH by distribution techniques is based on the estimate of the intracellular or intraorganellar concentration of the pH-indicating substance, but the usual techniques yield a value of the total amount of the indicator present, e.g., in the supernatant, pellet, filtrate, or filter. To obtain the intracellular concentration the total intracellular indicator value and the intracellular water volume must be calculated. The distribution method, therefore, involves the determination of intra- and extracellular water.

A number of techniques are known to achieve this.[88,89] One usually determines the concentration of a labeled or colored substance not penetrating the cell or organelle (or cells in a piece of tissue) and, in parallel, the concentration of a substance that does penetrate the cell or organelle. Comparison with the concentration before adding the cells or tissue shows directly how much extracellular water has been added to the testing solution, in the first case, and how much extracellular plus intracellular water has been added, in the second case.

The use of tritiated water as a membrane-permeant marker and a radiolabeled membrane-impermeant solute seems to be the most convenient approach, namely, the triple isotope method[21] employing a ^{14}C-labeled pH-indicating substance, $^{3}H_2O$ and $^{36}Cl^-$ or $^{35}SO_4^{2-}$ (thus, minimizing the experimental error by making calculations dependent only on ratios of the counts in the channels). Instead of chloride or sulfate, ^{14}C-labeled dextran, inulin, sorbitol, or sucrose is used for the determination of extracellular space. Alternatively, spin-labeled substances[74] (with consequent quenching of the signal from extracellularly located marker molecules by a membrane-impermeant quenching substance), NMR technique,[84,85] direct chemical methods, or determination of water by drying may be used for the determination of the intracellular volume.

The accuracy of the estimates of intracellular volume depends, however, on the marker employed as, e.g., inulin tends to underestimate and chloride to overestimate the measured value. Furthermore, cells and organelles may exhibit an osmotic response in the presence of the marker.

B. Temporal and Spatial Resolution

1. Temporal Resolution

The temporal resolution of distribution techniques is principally limited by the time needed for the equilibration of the distributive indicator between the intracellular and extracellular space. If a slowly responding technique for monitoring the indicator concentration is used, the overall time resolution is further decreased.

The time required for the equilibration depends on the indicator used and on the surface-to-volume ratio.[90] For instance, with DMO it ranges between 30 s for small microorganisms or organelles and several hours for larger cells in suspensions, tissues, or whole organs.[18]

The distribution of the indicator may be followed almost instantly if optical absorption and fluorescence techniques are used, while in the case of magnetic resonance techniques, considerable time may be required for the accumulation of the spectra.

The response time of ion-selective electrodes is of the order of magnitude of 1 to 10 s.

Monitoring of radiolabeled indicators may become relatively time-consuming, as it includes centrifugation or filtration of cells. The measurement is usually limited to one per sample. The introduction of flow dialysis represents a very elegant solution, allowing continuous monitoring of the indicator uptake with a time constant of the apparatus between 30 s and 10 min.[11,12]

2. Spatial Resolution

All the techniques give an average reading of pH values of all cells in the sample. However, radiolabeled indicators could provide individual cell pH readings or possibly even intracellular differentiation in pH values if autoradiography is used. Indeed, a photometric densitometry of autoradiographic images of teeth with ^{14}C-labeled DMO has demonstrated this possibility by showing differences in pH between various cell layers of dental soft tissues.[16]

C. Problems and Sources of Error in the Distribution Techniques

A distribution technique may fail if the distributive indicator does not exhibit the required properties.

The neutral form of the ideal distributive indicator should easily permeate the membrane, whereas the charged form should be fully membrane impermeant. The indicator should be passively distributed across the membrane without a contribution of any active transport process, any binding to the membrane, cell wall, or intracellular constituents, and it should not be metabolized. The most likely sources of errors are the following:[4,18,90]

1. Non-negligible permeability of the ionized form of the indicator
2. Metabolism of the indicator
3. Binding of the indicator to cell structures
4. Active transport of the indicator
5. Low permeability of the non-ionized form of the indicator
6. Erroneous estimates of the intracellular volume
7. Differences in p*K* of the indicator between intracellular and extracellular space
8. Direct effect of the presence of the indicator on the measured intracellular pH value

First, the permeability of the ionized form of a weak electrolyte is a potentially serious source of error. The higher the ratio of permeabilities of the charged form relative to the uncharged one and the higher the proportion of the charged form present (given by the indicator p*K* and ambient pH), the higher is the error introduced in the pH_{in} estimation.

The permeability of the ionized form of most indicators (DMO, trimethylamine, morphine) is, indeed, very low, resulting in under- or overestimation of the true pH only by some 0.01 pH units (e.g., DMO, ratio 1:1000, pH_{in} and pH_{out} within 1 to 2 pH units of the p*K* ~6.2).[14,91,92] On the other hand, ionic forms of methylamine and ammonia were found to be highly permeant in a number of cells. This permeability, together with high values leading to a high proportion of the charged form, may represent a significant threat to intracellular pH measurement. A detailed discussion of these and related problems may be found in the papers by Roos and Boron.[4,92-94]

Second, one of the criteria for choosing a suitable pH indicator is to prove that it is not metabolized. The commonly used DMO is only metabolized in a few plant cells.[95] The weak bases nicotine, morphine, atropine, phenobarbital, and trimethylamine are also probably not metabolized, though nicotine may be metabolized in some mammalian cells.[4] Bromophenol blue is definitely not metabolized.[1]

Third, the binding of the indicator inside the cell and on the cell surface may seriously distort the measurement. Its effect becomes less apparent if high concentrations of the indicator are employed so that the binding sites are saturated and most of the indicator remains free. On the other hand, the high concentration of an indicator increases the danger of altering the measured intracellular pH by the very presence of the indicator.

Though excessive binding may invalidate the measurement, reversible binding may still allow qualitative or semiquantitative estimates of the intracellular pH values. Indeed, in many biological experiments, a direct quantification of the data is not always necessary.

In the usual range of concentrations, DMO appears not to be bound in either animal or plant cells,[4] while this may not be true for nicotine.[38]

Fourth, the active transport of the indicator perturbs the equilibrum difference in concentrations established by the passive permeation of the uncharged form of the indicator. With a reasoning similar to that in the case of binding, this effect should become negligible at levels well above the transport-saturating concentration. Active transport of DMO has been demonstrated in some cells.[18,96-98]

Fifth, low permeability of the uncharged form of the indicator may practically never allow the true equilibrium distribution to be reached. This condition is highly vulnerable to various perturbations, such as the above-mentioned effect of the active transport of the indicator.

Sixth, the estimated value of the intracellular volume depends on the marker used. The true value may be both overestimated (e.g., with chloride) or underestimated (e.g., with inulin). Osmotic effects may also appear and distort the measurement. A quantitative assessment of errors arising from the use of an unsuitable marker was done by Robson et al.[17] It appears that the derived value of intracellular pH is usually rather insensitive to differences in the estimates of intracellular volume. The effect may become significant only at very low distribution ratios of the indicator (i.e., if $pH_{in} \ll pH_{out}$ for weak acids or $pH_{in} \gg pH_{out}$ for weak bases). Nevertheless, the differences in the value of the intracellular volume affect only absolute values of the measured intracellular pH and not changes in pH between parallel measurements of the same object under different experimental conditions.

Seventh, the formulae for the determination of the intracellular pH (Equations 91 to 93) are derived from the assumption that there is no difference in the p*K* value of the indicator between the intracellular and extracellular space. It can be shown that the numerical error in the calculation of intracellular pH is of the same magnitude as the difference in p*K*.[17] The stability of p*K* is thoroughly discussed in Chapter 4A.

Last of all, the measurement procedure may affect the intracellular pH because the simple suspension of cells in a solution of a weak acid or base and the subsequent entry of the neutral form of the electrolyte may alter the intracellular pH. Weak acids decrease the pH_{in} value; weak bases increase it. The final effect depends on the amount of indicator used and the buffering capacity of the cell.[99,100]

D. Examples of Distribution Technique Uses

A few examples from the literature will now be given to illustrate the use of distribution techniques, with an emphasis on work with microorganisms since this has been intentionally omitted from all the otherwise excellent reviews on intracellular pH.[4,18,101]

Bromophenol blue has been used to great advantage with several species of yeast.[1,55,56] It was found in the first systematic study of this type that intracellular pH at an "average" value, to be sure, was well buffered in two ranges of pH_{out} (Figure 14) with a transition at about pH 6.5. In the study, the effect of adding an acid-producing substrate, such as D-glucose, was described (Figure 15) and later confirmed.[56]

Propionic acid was used with both bacteria and yeasts in more recent studies,[44,102] and the pH was shown to be nearly a mirror image of the $\Delta\phi$ (the transmembrane potential). It appears to be a common observation with various cell types that the protonmotive force does not change much with external pH, this being due to the opposite behavior of Δ pH and $\Delta\phi$ (Figure 16).[102]

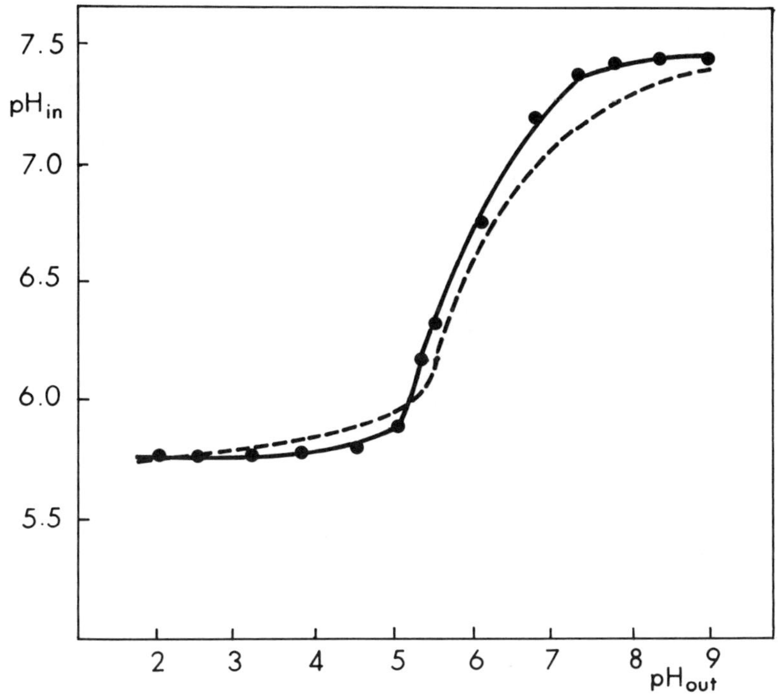

FIGURE 14. Dependence of intracellular on extracellular pH using the method of distribution of bromophenol blue in the yeast *Saccharomyces cerevisiae*. Temperature was 25°C. Buffers used were phthalate-HCl for pH = 2.1 to 3.8, phthalate-NaOH for pH = 4.6 to 6.9, and *Tris*-HCl for pH = 7.2 to 9.0. The broken curve shows the results of measurement of intracellular pH of this yeast using the dual-wavelength fluorescence technique with fluorescein, in 0.1 *M* triethanolamine-phthalic acid. (Values taken from Reference 1.)

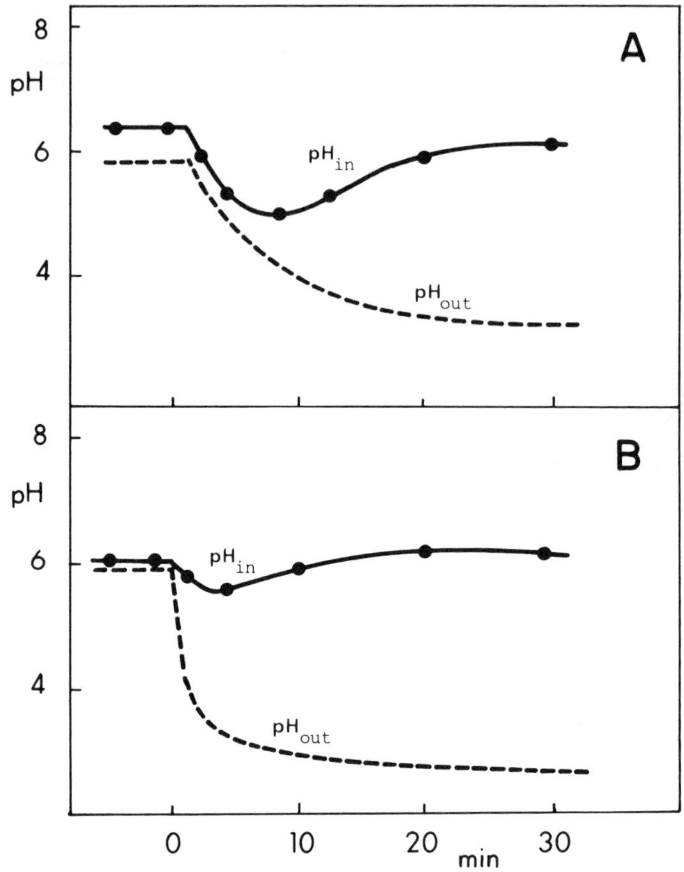

FIGURE 15. Intracellular and extracellular pH values of the yeast *Saccharomyces cerevisiae*, incubated aerobically at 30°C in an unbuffered medium. At time 0, D-glucose was added to 0.1 *M*. The intracellular pH was estimated from the distribution of bromophenol blue; the extracellular pH was recorded by a conventional glass electrode. The suspension density in (A) was 15 mg dry weight per milliliter; in (B) it was 75 mg dry weight per milliliter. (Data from Reference 23.)

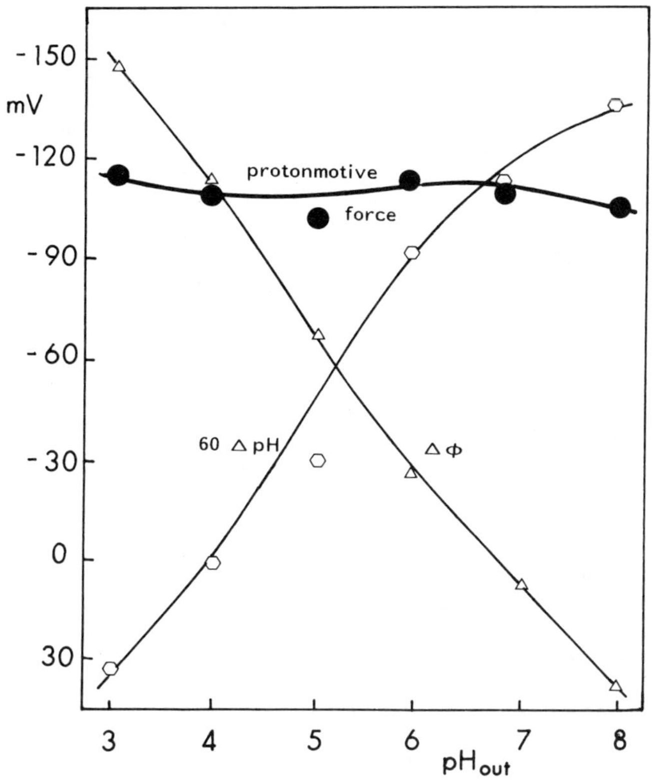

FIGURE 16. pH dependence of the components of protonmotive force (pmf = $\Delta\phi - 60\Delta$pH) in the yeast *Rhodotorula gracilis* at 30°C. (Data from Reference 22 and from this laboratory.)

REFERENCES

1. **Kotyk. A.,** Intracellular pH of baker's yeast, *Folia Microbiol.,* 8, 27, 1963.
2. **Rottenberg, H.,** The measurement of membrane potential and Δ pH in cells, organelles, and vesicles, *Methods Enzymol.,* 55, 547, 1979.
3. **Kotyk, A.,** pH effects on membrane transport, in *Structure and Properties of Cell Membranes,* Vol. 2, Benga, Gh., Ed., CRC Press, Boca Raton, FL, 1985, 11.
4. **Roos, A. and Boron, W. F.,** Intracellular pH, *Physiol. Rev.,* 61, 296, 1981.
5. **Wadell, W. J. and Bates, R. G.,** Intracellular pH, *Physiol. Rev.,* 49, 285, 1969.
6. **Adler, S.,** The simultaneous determination of muscle cell pH using a weak acid and a weak base, *J. Clin. Invest.,* 51, 256, 1972.
7. **Peschek, G. A., Nitschmann, W. H., and Czerny, T.,** Respiratory proton extrusion and plasma membrane energization, *Methods Enzymol.,* 167, 367, 1988.
8. **Nitschmann, W. H.,** Chemiosmotic coupling in cyanobacteria — ATP synthesis in a two-membrane system, *J. Theor. Biol.,* 122, 409, 1986.
9. **Falkner, G., Horner, F., Werdan, K., and Heldt, H. W.,** pH changes in the cytoplasm of the blue-green alga Anacystis nidulans caused by light-dependent proton flux into the thylakoid space, *Plant Physiol.,* 58, 717, 1976.
10. **Heldt, H. W., Werdan, K., Milovancev, M., and Geller, G.,** Alkalization of the chloroplast stroma caused by light-dependent proton flux into thylakoid space, *Biochim. Biophys. Acta,* 314, 224, 1973.
11. **Ramos, S., Schuldiner, S., and Kaback, H. R.,** The electrochemical gradient of protons and its relationship to active transport in Escherichia coli membrane vesicles, *Proc. Natl. Acad. Sci. U.S.A.,* 73, 1892, 1986.

12. **Ramos, S., Schuldiner, S., and Kaback, H. R.,** The use of flow dialysis for determination of Δ pH and active transport, *Methods Enzymol.*, 55, 680, 1979.
13. **Waddell, W. J. and Butler, T. C.,** Calculation of intracellular pH from the distribution of 5,5-dimethyl-2,4-oxazolidinedione (DMO). Application to skeletal muscle of the dog, *J. Clin. Invest.*, 38, 720, 1959.
14. **Boron, W. F. and Roos, A.,** Comparison of microelectrode, DMO and methylamine methods for measuring intracellular pH, *Am. J. Physiol.*, 231, 799, 1976.
15. **Hinke, J. A. M. and Menard, M. R.,** Intracellular pH of single crustacean muscle fibres by the DMO and electrode methods during acid and alklaline conditions, *J. Physiol. (London)*, 262, 533, 1976.
16. **Lyman, G. E. and Waddell, W. J.,** pH gradients in the developing teeth of young mice from autoradiography of (^{14}CO)DMO, *Am. J. Physiol.*, 232, F364, 1977.
17. **Robson, J. S., Bone, J. M., and Lambie, A. T.,** Intracellular pH, *Adv. Clin. Chem.*, 213, 1968.
18. **Cohen, R. D. and Iles, R. A.,** Intracellular pH: measurement, control, and metabolic interrelationships. *CRC Crit. Rev.*, 6, 101, 1975.
19. **Hinke, J. A. M. and Menard, M. R.,** Evaluation of the DMO method for measuring intracellular pH, *Respir. Physiol.*, 33, 31, 1978.
20. **Scholz, A. and Baethke, R.,** Assessment of the intracellular pH value with carbon-14 labeled 5,5-dimethyl-2,4-oxazolidinedione, *Klin. Wochenschr.*, 49, 994, 1971.
21. **Schloerb, P. R. and Grantham, J. J.,** Intracellular pH measurement with tritiated water, carbon-14 labelled 5,5-dimethyl-2,4-oxazolidinedione, and chloride-36, *J. Lab. Clin. Med.*, 65, 669, 1965.
22. **Schummer, U., Schiefer, H. G., and Gerhardt, U.,** The proton gradient across mycoplasma membranes, *Curr. Microbiol.*, 5, 371, 1981.
23. **Granholm, L., Kaasik, A. E., Nilsson, L., and Siesjö, B. K.,** The lactate/pyruvate ratios of cerebrospinal fluid of rats and cats related to the lactate/pyruvate, the ATP/ADP, and the phosphocreatine/creatine ratios of brain tissue, *Acta Physiol. Scand.*, 74, 398, 1968.
24. **Granholm, L. and Siesjö, B. K.,** The effects of hypercapnia and hypocapnia upon cerebrospinal fluid lactate and pyruvate concentrations and upon the lactate, pyruvate, ATP, ADP, phosphocreatine, and creatine concentrations of cat brain tissue, *Acta Physiol. Scand.*, 75, 257, 1969.
25. **Roos, A.,** Intracellular pH and distribution of weak acids across cell membranes. A study of D- and L-lactate and of DMO in rat diaphragm, *J. Physiol. (London)*, 249, 1, 1975.
26. **Bassnett, S., Croghan, P. C., and Duncan, G.,** Diffusion of lactate and its role in determining intracellular pH in the lens of the eye, *Exp. Eye Res.*, 44, 143, 1987.
27. **Rotterberg, H.,** The mechanism of energy-dependent ion transport in mitochondria, *J. Theor. Biol.*, 11, 117, 1973.
28. **De la Peña, P., Barros, F., Gascón, S., Ramos, S., and Lazo, P. S.,** The electrochemical proton gradient of Saccharomyces. The role of potassium, *Eur. J. Biochem.*, 123, 447, 1982.
29. **Seaston, A. G., Carr, G., and Eddy, A. A.,** The concentration of glycine by preparations of the yeast Saccharomyces carlsbergensis depleted of adenosine triphosphate, *Biochem. J.*, 154, 669, 1976.
30. **Aerts, R. J., Durston, A. J., and Moolenaar, W. H.,** Cytoplasmic pH and glycolysis in the Dictyostelium discoideum cell cycle, *FEBS Lett.*, 196, 167, 1986.
31. **Garcia-Sancho, J. and Sanchez, A.,** Use of salicylic acid to measure the apparent intracellular pH in the Ehrlich ascites-tumor cell and Escherichia coli, *Biochim. Biophys. Acta*, 509, 148, 1978.
32. **Slayman, C. L., and Slayman, C. W.,** Net uptake of potassium in Neurospora. Exchange for sodium and hydrogen ions, *J. Gen. Physiol.*, 52, 424, 1968.
33. **Milligan, C. L. and Wood, C. M.,** Intracellular pH transients in rainbow trout tissues measured by dimethadione distribution, *Am. J. Physiol.*, 248, R668, 1985.
34. **Brown, D. A. and Garthwaite, J.,** Intracellular pH and the distribution of weak acids and bases in isolated rat superior cervical ganglia, *J. Physiol. (London)*, 297, 597, 1979.
35. **Friderici, L. S.,** Exchange of chloride ions and of carbon dioxide between blood corpuscles and blood plasma, *J. Biol. Chem.*, 42, 245, 1920.
36. **Siesjö, B. K., Folbergrová, J., and Macmillan, V.,** Effect of hypercapnia upon intracellular pH in the brain, evaluated by the bicarbonate-carbonic acid method and from the creatine phosphokinase equilibrium, *J. Neurochem.*, 19, 2483, 1972.
37. **Fenn, W. O. and Maurer, F. W.,** The pH of muscle, *Protoplasma*, 24, 337, 1935.
38. **Weiss, G. B.,** Dependence of nictotine-^{14}C distribution and movements upon pH in from sartorius muscle, *J. Pharmacol. Exp. Ther.*, 160, 135, 1968.
39. **Hudgins, P. M. and Putney, J. W.,** Distribution of local anesthetics and the intracellular pH in vascular smooth muscle, *J. Pharmacol. Exp. Ther.*, 181, 538, 1972.
40. **Hannan, S. F. and Wiggins, P. M.,** Intracellular pH of frog sartorius muscle, *Biochim. Biophys. Acta*, 428, 205, 1976.
41. **Brown, D. A. and Garthwaite, J.,** Intracellular pH and the distribution of weak acids and bases in isolated rat superior cervical ganglia, *J. Physiol. (London)*, 297, 597, 1979.

42. **Goldman, R. and Rottenberg, H.,** Ion distribution in lysosomal suspensions, *FEBS Lett.*, 33, 233, 1973.
43. **Rottenberg, H., Grunwald, T., and Avron, M.,** Determination of Δ pH in chloroplasts. I. Distribution of (^{14}C)methyl amine, *Eur. J. Biochem.*, 25, 54, 1972.
44. **Rottenberg, H., Grunwald, T., and Avron, M.,** Direct determination of Δ pH in chloroplasts, and its relation to the mechanisms of photoinduced reactions, *FEBS Lett.*, 13, 41, 1971.
45. **Lee, H. C.,** Sodium and proton transport in flagella isolated from sea urchin spermatozoa, *J. Biol. Chem.*, 259, 4957, 1984.
46. **Gaensslen, R. E. and McCarty, R. E.,** Amine uptake in chloroplasts, *Arch. Biochem. Biophys.*, 147, 55, 1971.
47. **Netter, H.,** Gehorcht die Ammoniak-Verteilung auf Blutkörperchen und Serum den Membrangleichgewichten?, *Pflügers Arch.*, 222, 724, 1929.
48. **Netter, H.,** Die Stellung des Kaliums in der Elektrolytsystem des Muskels, *Pflügers Arch.*, 234, 680, 1934.
49. **Fenn, W. O., Haege, L. F., Sheridan, E., and Flick, J. B.,** The penetration of ammonia into frog muscle, *J. Gen. Physiol.*, 28, 53, 1944.
50. **Bone, J. M., Verth, A., and Lambie, A. F.,** Intracellular acid-base heterogeneity, in *Proc. 9th Annu. Meet. Eur. Soc. Clin. Invest.*, Rotterdam, 1975, 24.
51. **Robin, E. D., Vester, J. W., Wilson, R. J., and Andrus, M. H.,** The "internal" pH of mitochondria with observations on the functional significance of mitochondrial membranes, *J. Clin. Invest.*, 39, 1022, 1960.
52. **Rottenberg, H. and Grunwald, T.,** Determination of Δ pH in chloroplasts. III. Ammonium uptake as a measure of Δ pH in chloroplasts and sub-chloroplast particles, *Eur. J. Biochem.*, 25, 71, 1972.
53. **Cirillo, V. P. and Gromet-Elhanan, Z.,** Steady-state measurements of Δ pH and Δ Ψ in Rhodospirillum rubrum chromatophores by two different methods. Comparison with phosphorylation potential, *Biochim. Biophys. Acta*, 636, 244, 1981.
54. **Gräber, P. and Witt, H. T.,** Relations between the electrical potential, pH gradient, proton flux and phosphorylation in the photosynthetic membrane, *Biochim. Biophys. Acta*, 423, 141, 1976.
55. **Höfer, M. and Misra, P. C.,** Evidence for a proton/sugar symport in the yeast Rhodotorula gracilis (glutinis), *Biochem. J.*, 172, 15, 1978.
56. **Sigler, K., Knotková, A., and Kotyk, A.,** Factors governing substrate-induced generation and extrusion of protons in the yeast Saccharomyces cerevisiae, *Biochim. Biophys. Acta*, 643, 572, 1981.
57. **Schuldiner, S., Rottenberg, H., and Avron, M.,** Determination of Δ pH in chloroplasts. II. Fluorescent amines as a probe for determination of Δ pH in chloroplasts, *Eur. J. Biochem.*, 25, 64, 1972.
58. **Deamer, D. W., Prince, R. C., and Crofts, A. R.,** The response of fluorescent amines to pH gradients across liposome membranes, *Biochim. Biophys. Acta*, 274, 323, 1972.
59. **Fiolet, J. W. T., Bakker, E. P., and Van Dam, K.,** The fluorescent properties of acridines in the presence of chloroplasts or liposomes. On the quantitative relationship between the fluorescence quenching and the transmembrane proton gradient, *Biochim. Biophys. Acta*, 368, 432, 1974.
60. **Dencher, N. A., Burghaus, P. A., and Grzesiek, S.,** Determination of the net proton-hydroxide ion permeability across vesicular lipid bilayers and membrane proteins by optical probes, *Methods Enzymol.*, 127, 746, 1986.
61. **Schummer, U., Schiefer, H. G., and Gerhardt, U.,** The proton gradient across mycoplasma membranes, *Curr. Microbiol.*, 5, 371, 1981.
62. **Elema, R. P., Michels, P. A. M., and Konings, W. N.,** Response of 9-aminoacridine fluorescence in transmembrane pH-gradients in chromatophores from Rhodopseudomonas sphaeroides, *Eur. J. Biochem.*, 92, 381, 1978.
63. **Lee, H. C. and Forte, J. G.,** A study of H^+ transport in gastric microsomal vesicles using fluorescent probes, *Biochim. Biophys. Acta*, 508, 339, 1978.
64. **Casadio, R. and Melandri, B. A.,** The behavior of 9-aminoacridine as an indicator of transmembrane pH difference in liposomes of natural bacterial phospholipids, *J. Bioenerg. Biomembr.*, 9, 17, 1977.
65. **Casadio, R., Baccarini-Melandri, A., and Melandri, B. A.,** On the determination of the transmembrane pH difference in bacterial chromatophores using 9-aminoacridine, *Eur. J. Biochem.*, 47, 121, 1974.
66. **Searle, G. F. W. and Barber, J.,** The involvement of the electrical double layer in the quenching of 9-aminoacridine fluorescence by negatively charged surfaces, *Biochim. Biophys. Acta*, 502, 309, 1978.
67. **Salama, G., Johnson, R. G., and Scarpa, A.,** Spectrophotometric measurements of transmembrane potential and pH gradients in chromaffin granules, *J. Gen. Physiol.*, 75, 109, 1980.
68. **Haraux, F. and de Kouchovski, Y.,** Measurement of chloroplast internal protons with 9-aminoacridine. Probe binding, dark proton gradient, and salt effects, *Biochim. Biophys. Acta*, 592, 153, 168.
69. **Singh, A. P. and Bragg, P. D.,** ATP-dependent proton translocation and quenching of 9-aminoacridine fluorescence in inside-out membrane vesicles of a cytochrome-deficient mutant of Escherichia coli, *Biochim. Biophys. Acta*, 464, 562, 1977.

70. **Lee, H. C., Forte, J. G., and Epel, D.,** The use of fluorescent amines for the measurement of pH_i, in *Intracellular pH: Its Measurement, Regulation and Utilization in Cellular Functions,* Nuccitelli, R. and Deamer, D. W., Eds., Alan R. Liss, New York, 1982, 135.
71. **Hinterstoisser, B. and Peschek, G. A.,** Fluorometric pH measurement in whole cells of dark aerobic and anaerobic cyanobacteria, *FEBS Lett.,* 217, 169, 1987.
72. **Gromet-Elhanan, Z.,** Relationship between light-induced quenching of atebrin fluorescence and ATP formation in Rhodospirillum rubrum chromatophores, *FEBS Lett.,* 13, 124, 1971.
73. **Lee, H. C. and Forte, J. G.,** A study of H^+ transport in gastric microsomal vesicles using fluorescent probes, *Biochim. Biophys. Acta,* 508, 339, 1978.
74. **Belkin, S., Melhorn, R. J., and Packer, L.,** Proton gradients in intact cyanobacteria, *Plant Physiol.,* 84, 25, 1987.
75. **Mehlhorn, R. J., Candau, P., and Packer, L.,** Measurements of volume and electrochemical gradients with spin probes in membrane vesicles, *Methods Enzymol.,* 88, 751, 1982.
76. **Mehlhorn, R. J. and Probst, I.,** Light-induced pH gradients measured with spin-labeled amine and carboxylic acid probes: application to Halobacterium halobium cell envelope vesicles, *Methods Enzymol.,* 88, 334, 1982.
77. **Kamo, N., Takeuchi, M., Hazemoto, N., and Kobatake, Y.,** Light-induced ΔpH of envelope vesicles containing halorodopsin measured by use of a spin probe, *Arch. Biochem. Biophys.,* 221, 514, 1983.
78. **Melandri, B. A., Mehlhorn, R. J., and Packer, L.,** Light induced proton gradients and internal volumes in chromatophores of Rhodopseudomonas sphaeroides, *Arch. Biochem. Biophys.,* 235, 97, 1984.
79. Molecular Probes, Inc., *Handbook of Spin Labels,* Junction City, OR, 1984.
80. **Wille, B. and Lavergne, J.,** Measurement of proton translocation in thylakoids under flashing light using a spin-labeled amine, *Photochem. Photobiophys.,* 4, 131, 1982.
81. **Nesbitt, D. M. and Berg, S. P.,** Proton involvement with the light-induced hindrances of spin label motion in the lumen of spinach thylakoids, *Biochim. Biophys. Acta,* 593, 353, 1980.
82. **Quantanilha, A. T. and Mehlhorn, R. J.,** pH gradients across thylakoid membranes measured with a spin-labelled amine, *FEBS Lett.,* 91, 104, 1978.
83. **Mehlhorn, R. J. and Packer, L.,** Bioenergetic studies of cells with spin probes, *Ann. N.Y. Acad. Sci.,* 414, 180, 1983.
84. **Hoffman, D., Kumar, A. M., Spitze, A., and Gupta, R. K.,** NMR measurement of intracellular water volume in rat kidney proximal tubules, *Biochim. Biophys. Acta,* 889, 355, 1986.
85. **Kirk, K. and Kuchel, P. W.,** Red cell volume changes monitored using a new ^{31}P NMR procedure, *J. Magn. Reson.,* 62, 568, 1985.
86. **Stewart, I. M., Chapman, B. E., Kirk, K., Kuchel, P. W., Lovric, V. A., and Raftos, J. E.,** Intracellular pH in stored erythrocytes. Refinement and further characterization of the ^{31}P NMR methylphosphonate procedure, *Biochim. Biophys. Acta,* 885, 22, 1986.
87. **Raftos, J. E., Chapman, B. E., Kuchel, P. W., Lovric, V. A., and Stewart, I. M.,** Intra- and extraerythrocyte pH at 37°C and during long-term storage at 4°C: ^{31}P NMR measurements and an electrochemical model of the system, *Haematologia,* 19, 21, 1986.
88. **Kotyk, A. and Janáček, K.,** *Cell Membrane Transport,* Plenum Press, New York, 1975.
89. **Dick, D. A. T.,** *Cell Water,* Butterworths, London, 1966.
90. **Roos, A. and Keifer, D. W.,** Estimation of intracellular pH from distribution of weak electrolytes, in *Intracellular pH: Its Measurement, Regulation and Utilization in Cellular Functions,* Nuccitelli, R. and Deamer, D. W., Eds., Alan R. Liss, New York, 1982, 55.
91. **Keifer, D. W. and Roos, A.,** Membrane permeability to the molecular and ionic forms of DMO in barnacle muscle, *Am. J. Physiol.,* 240, C73, 1981.
92. **Roos, A.,** Intracellular pH and intracellular buffering power of the cat brain, *Am. J. Physiol.,* 209, 1233, 1965.
93. **Roos, A.,** Intracellular pH and buffering power of rat muscle, *Am. J. Physiol.,* 221, 182, 1971.
94. **Roos, A. and Boron, W. F.,** Intracellular pH transients in rat diaphragm muscle measured with DMO, *Am. J. Physiol.,* 235, C49, 1978.
95. **Kurkdjian, A. and Guern, J.,** Intracellular pH in higher plant cells. I. Improvements in the use of the 5,5-dimethyloxazolidine-2(^{14}C),4-dione distribution technique, *Plant Sci. Lett.,* 11, 337, 1978.
96. **Dietschy, J. M. and Carter, N. W.,** Active transport of 5,5-dimethyl-2,4-oxazolidinedione, *Science,* 150, 1294, 1965.
97. **Butler, T. C.,** Active transport of 5,5-dimethyl-2,4-oxazolidinedione, *Science,* 151, 1564, 1966.
98. **Dietschy, J. M. and Carter, N. W.,** Active transport of 5,5-dimethyl-2,4-oxazolidinedione, *Science,* 152, 799, 1966.
99. **Cole, M. B. and Keenan, M. H. J.,** Effects of weak acids and external pH on the intracellular pH of Zygosaccharomyces bailii, and its implication in weak-acid resistance, *Yeast,* 3, 23, 1987.

100. **Keifer, D. W.,** Effect of weak acids on pH regulation and anion transport in barnacle muscle fibers, *Am. J. Physiol.,* 241, C193, 1981.
101. **Nuccitelli, R. C. and Deamer, D. W.,** *Intracellular pH: Its Measurement, Regulation and Utilization in Cellular Functions,* Alan R. Liss, New York, 1982.
102. **Horák, J. and Kotyk, A.,** Energetics of L-proline uptake by Saccharomyces cerevisiae, *Biochim. Biophys. Acta,* 857, 173, 1986.

Chapter 4C

ASSAYS OF INTRACELLULAR pH USING CHEMICAL PROBES: ABSORPTION SPECTROSCOPY

Jan Slavík

TABLE OF CONTENTS

I. THEORETICAL ASPECTS

A. Introduction

Absorption spectroscopy is a widely used, "old faithful" technique. If visual colorimetry by the naked eye can be accepted as an absorption measurement, the first intracellular pH measurement dates back as far as 1871. In the first experiments, naturally occurring intracellular anthocyanine pigments were employed. Later, extrinsic dyes, such as litmus, were brought inside the cell by digestion, injection, or incubation in solution. However, the colorimetric measurements were abandoned after 1930 in favor of other, mainly distribution, techniques. Novel applications appeared as late as in the 1970s, some of them taking advantage of high temporal and spatial resolution (topography).

The main advantage of absorption techniques probably lies in their clarity and simplicity. The principles are easy to grasp, measurement is uncomplicated, straightforward, and fast. The instruments are available at reasonable prices. Cells can be visually observed during the measurement, either directly or under an optical microscope.

The weakest point may be the low sensitivity (as compared with fluorescence), requiring the use of higher intracellular concentrations of the dye.

The protocol of an absorption experiment requires little experience and physical knowledge and can be easily done by a single person (in contrast with NMR).

B. The Basis of Color

Selective absorption of light results in the complementary color of the absorbing sample. For instance, a blue-absorbing compound appears yellow in both transmitted and reflected light.

The absorption properties in the near-ultraviolet and visible light ranges (200 to 800 nm) are roughly attributable to specific chemical groups (e.g., =C=O, –N=N–, –N=O, –C=O, $-NO_2$, –N≡N, or aromatic and conjugated systems) and to chemical groups in their close neighborhood. The whole chromophoric unit behaves independently and may be, as a so-called absorption (or fluorescent) "label" or "marker", attached to another submolecular unit or molecule. If there are two chromophores in one molecule, the absorption spectrum of this molecule represents a superposition of two independent absorption bands. Larger chromophoric systems generally show broader and more intense absorption bands.

Visible absorption is usually caused by π-electron transitions ($n \rightarrow \pi^*$ or $\pi \rightarrow \pi^*$). Saturated hydrocarbon chains do not show visible absorption because the $\sigma \rightarrow \sigma^*$ transition corresponds to 135 nm.

Unfortunately, prediction of the shape and spectral position of an absorption band based on the knowledge of a chemical formula is difficult. For instance, fluorescein is red as a powder, yellow in solution, and colorless if acetated.

C. The Phenomenon of Absorption

Electrons occupy the states with lowest energy among the large number of possible states allowed by quantum-mechanical rules (Figure 17). However, if energy is received from the outside (e.g., absorption of a photon that hits the molecule), one of the electrons moves to a higher state (this process takes about 1 fs) and, after a few ps or ns returns to its original state. (The return of the electron to the ground state may be accompanied by the release of a fluorescence photon.)

However, the absorption of light occurs only if the energy of the light equals the energy difference between the two electron states, or

$$\Delta E = E_1 - E_0 \tag{97a}$$

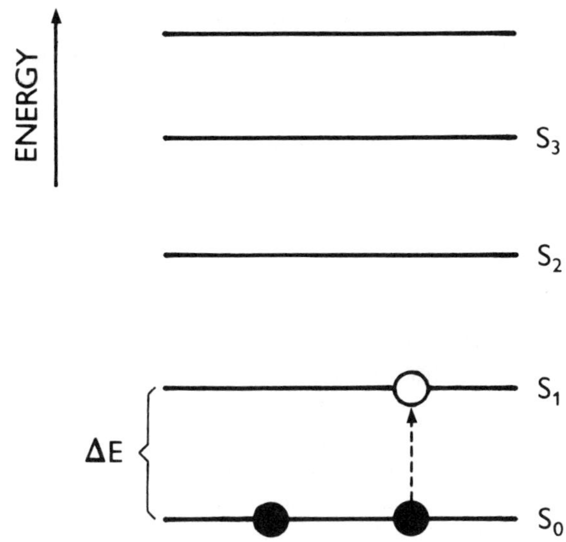

FIGURE 17. Occupancy of electron energy levels and the process of light absorption.

Using the relation between energy, frequency, and wavelength of light $\Delta E = h\nu = hc/\lambda$ (h is Planck's constant and c is the velocity of light), Equation 97a may be expressed in terms of wavelength

$$\lambda = hc/(E_1 - E_o) \tag{97b}$$

According to these equations, only light of a particular wavelength, exactly corresponding to ΔE, may be absorbed. In a complex molecule, however, light can be absorbed also if its wavelength does not correspond to Equation 97b. The small energy differences may be balanced at the expense of vibrational and rotational motion of the molecules. Therefore, in practice, broad absorption bands instead of narrow peaks are observed.

D. What Is Measured?

Absorption measurements are based on a comparison of the intensity of incident light (I_0) with the intensity of light that was passed through the sample (I). If c is the concentration of the dye and l the length of the light path, the following parameters can be defined:

Transparency	$T = I/I_0$ (usually given in %)
Absorbance (optical density)	$A = -\log T = \log (I_0/I)$
Absorptivity (Absorption coefficient)	$a = A/l = \log (I_0/I)/l$
Molar absorptivity (Molar extinction coefficient)	$\epsilon = A/cl = \log (I_0/I)/cl$

The typical values of molar absorptivity at maximum absorption wavelength are 500 to 100,000 for the $\pi \rightarrow \pi^*$ transitions and 10 to 1000 for the $n \rightarrow \pi^*$ transitions.

E. Basic Relations between Absorption Parameters

The basic equation relating the concentration of the dye to the observed attenuation of light is the Lambert-Beer Law

$$I = I_0 \, e^{-\epsilon c l} \tag{98a}$$

More conveniently, it is expressed as absorbance

$$A = \epsilon c l \tag{98b}$$

The Lambert-Beer Law applies to any solute of concentration below approximately 0.01 *M*, provided that molecules of the solute do not interact with each other (e.g., no dimerization occurs) and that the light is truly monochromatic.

Changes in both position and intensity of absorption bands due to the aggregation or dimer formation may be quite significant. It is known that a simple exchange of a polar solvent for a nonpolar one causes a shift in the maximum absorption wavelength of up to 3 to 40 nm, depending on the type of the electron transition ($n \rightarrow \pi^*$, $n \rightarrow \sigma^*$, $\pi \rightarrow \pi^*$). Chromophores with heteroatoms (C=O, N=O) are particularly sensitive.

If more than one solute is present, the Lambert-Beer Law reads

$$I = I_0 \, e^{-l \sum_{i=1}^{N} \epsilon_i c_i} \tag{99a}$$

or, in absorbance,

$$A = l \sum_{i=1}^{N} \epsilon_i c_i \tag{99b}$$

where N is the number of solutes.

The last equation is the basic equation of absorption measurements of intracellular pH by both single and double wavelength, as described below.

F. Theory of pH Measurement

The following text is based on the assumption that two differently colored species, A and B, exist. It shows how the ratio of concentrations of the two species R can be determined from an absorption measurement. In a formally similar form it can be applied to fluorescence measurements.

1. The Double-Wavelength Approach

Let us have two species, A and B, at concentrations denoted as c_A and c_B, with absorptivities $\epsilon^{A}_{\lambda_1}$ and $\epsilon^{B}_{\lambda_1}$ at wavelength λ_1, and $\epsilon^{A}_{\lambda_2}$ and $\epsilon^{B}_{\lambda_2}$ at wavelength λ_2.

According to the Lambert-Beer Law, the absorbances at λ_1 and λ_2 are

$$A_1 = \epsilon^{A}_{\lambda_1} c_A l + \epsilon^{B}_{\lambda_1} c_B l \tag{100}$$

$$A_2 = \epsilon^{A}_{\lambda_2} c_A l + \epsilon^{B}_{\lambda_2} c_B l \tag{101}$$

On dividing these two equations we obtain

$$A_1/A_2 = \frac{\epsilon^{A}_{\lambda_1} + \epsilon^{B}_{\lambda_1} R}{\epsilon^{A}_{\lambda_2} + \epsilon^{B}_{\lambda_2} R} \tag{102}$$

where $R = c_B/c_A$. Therefore, R is given by the ratio A_1/A_2 and by a set of constants characteristic for a particular dye (molar absorptivities). It is independent of the total dye concentration $c = c_A + c_B$ and of the optical path length. As an example, let us calculate R from Equation 102 and insert this value into Equation 81

$$\mathrm{pH} = \mathrm{p}K' + \log \frac{-\epsilon_{\lambda_1}^{A} + \epsilon_{\lambda_1}^{B}(A_1/A_2)}{\epsilon_{\lambda_2}^{A} - \epsilon_{\lambda_2}^{B}(A_1/A_2)} \tag{103}$$

Indeed, the calibration curve is independent of all instrumental constants and involves only the characteristics of the dye (the pK' value and the tetrad of absorptivities characteristic for the dye) and the measured ratio A_1/A_2.

It is clear from Equation 103 that to increase the accuracy of the measurement the two wavelengths should be selected such that

$$\epsilon_{\lambda_1}^{A} \ll \epsilon_{\lambda_1}^{B} \qquad \epsilon_{\lambda_2}^{A} \ll \epsilon_{\lambda_2}^{B} \tag{104}$$

In other words, λ_1 should be close to the absorption maximum of A and λ_2 to that of B. In practice, a comparison of the absorption spectra of both A and B is useful for finding the most suitable wavelengths for selective monitoring of A and B.

2. The Single-Wavelength Approach

Equations 100 and 101 contain two unknowns, c_A and c_B, that can be calculated

$$c_A = \frac{A_1\epsilon_{\lambda_2}^{B} - A_2\epsilon_{\lambda_1}^{B}}{l(\epsilon_{\lambda_1}^{A}\epsilon_{\lambda_2}^{B} - \epsilon_{\lambda_2}^{A}\epsilon_{\lambda_1}^{B})} \tag{105}$$

$$c_B = \frac{A_2\epsilon_{\lambda_1}^{B} - A_1\epsilon_{\lambda_2}^{A}}{l(\epsilon_{\lambda_1}^{A}\epsilon_{\lambda_2}^{B} - \epsilon_{\lambda_2}^{A}\epsilon_{\lambda_1}^{B})} \tag{106}$$

However, for the determination of pH it is only their ratio $R = c_B/c_A$ that is needed. Supposing that the total concentration of the dye is known and constant. Then, from the mathematical point of view, one measurement at one given wavelength is sufficient.

$$A = \epsilon^{A}c_{A}l + \epsilon^{B}c_{B}l \tag{107}$$

Instead of measuring the second wavelength, we resort to

$$c = c_A + c_B \tag{108}$$

and solve the two equations for c_A and c_B

$$R = c_B/c_A = \frac{\epsilon^{A} - \epsilon^{B}}{A/cl - \epsilon^{B}} - 1 \tag{109}$$

On inserting into Equation 78, we have

$$\mathrm{pH} = \mathrm{p}K' + \log\left(\frac{\epsilon^{A} - \epsilon^{B}}{A/cl - \epsilon^{B}} - 1\right) \tag{110}$$

Comparison of Equations 103 and 110 makes it clear that the single-wavelength measurement of pH requires a constant path length and a constant total concentration of the dye to be useful, and this makes the technique vulnerable. The results of routine comparative measurements should, thus, be occasionally confirmed by double-wavelength measurements. However, the single-wavelength approach is useful for rapid time-sequence measurements in devices where the simultaneous measurement at two wavelengths is technically not feasible.[1] In other cases, the double-wavelength measurements should be preferred to those of single wavelength. The intracellular topography of pH can be realized with double-wavelength measurements only (l and c vary).

A special case arises if the other form of the dye does not absorb light, i.e., $\epsilon^B = 0$ (e.g., phenolphthalein). Equation 107 then simplifies to

$$A = \epsilon^A c_A l \tag{111}$$

and, using Equation 108, we obtain by analogy with Equation 109,

$$R = \frac{\epsilon^A cl}{A} - 1 \tag{112}$$

and

$$\mathrm{pH} = \mathrm{p}K' + \log(\epsilon^A cl/A - 1) \tag{113}$$

G. Theory of the pH-Induced Color Change

Two theories attempt to explain the color change exhibited by a pH-sensitive dye. The older and simpler one is due to Ostwald, and the newer and more complex one was proposed by Hantzsch.

According to Ostwald, pH-indicating dyes are weak acids or bases which change their color upon dissociation. For instance, an acidobasic indicator HInd dissociates in water

$$\mathrm{HInd} + \mathrm{H_2O} \leftrightharpoons \mathrm{H_3O^+} + \mathrm{Ind^-} \tag{114}$$

with a dissociation constant pK'

$$K' = \frac{c_{\mathrm{H3O^+}}/c_{\mathrm{Ind^-}}}{c_{\mathrm{HInd}}} \tag{115}$$

Thus (Figure 18),

$$\mathrm{pH} = \mathrm{p}K' + \log(c_{\mathrm{Ind^-}}/c_{\mathrm{HInd}}) \tag{116}$$

If, according to the Ostwald theory, the dissociated and the undissociated forms of the dye differ in color, the ratio R, measured as described in the previous section, can be simply expressed as

$$R = c_{\mathrm{Ind^-}}/c_{\mathrm{HInd}} \tag{117}$$

and Equation 116 simplifies to (Figure 19)

$$\mathrm{pH} = \mathrm{p}K' + \log R \tag{118}$$

Unfortunately, the Ostwald theory is probably oversimplified. It seems unlikely that a mere dissociation of an acidic proton could cause such profound changes in the arrangement of the electron cloud manifested by a color change.

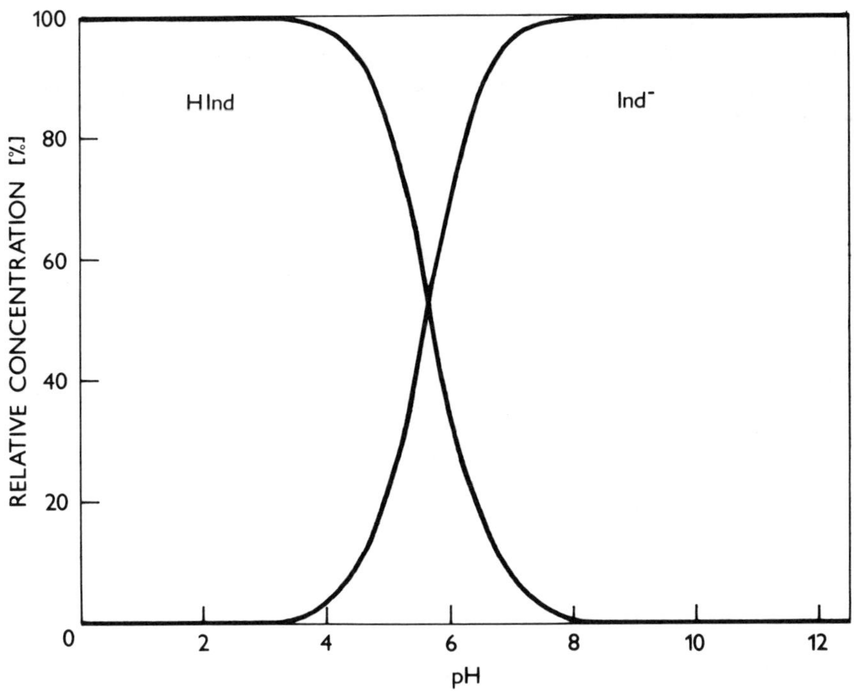

FIGURE 18. The relative concentrations of dissociated and undissociated forms of an indicator dye according to Ostwald's theory (pK' = 5.7).

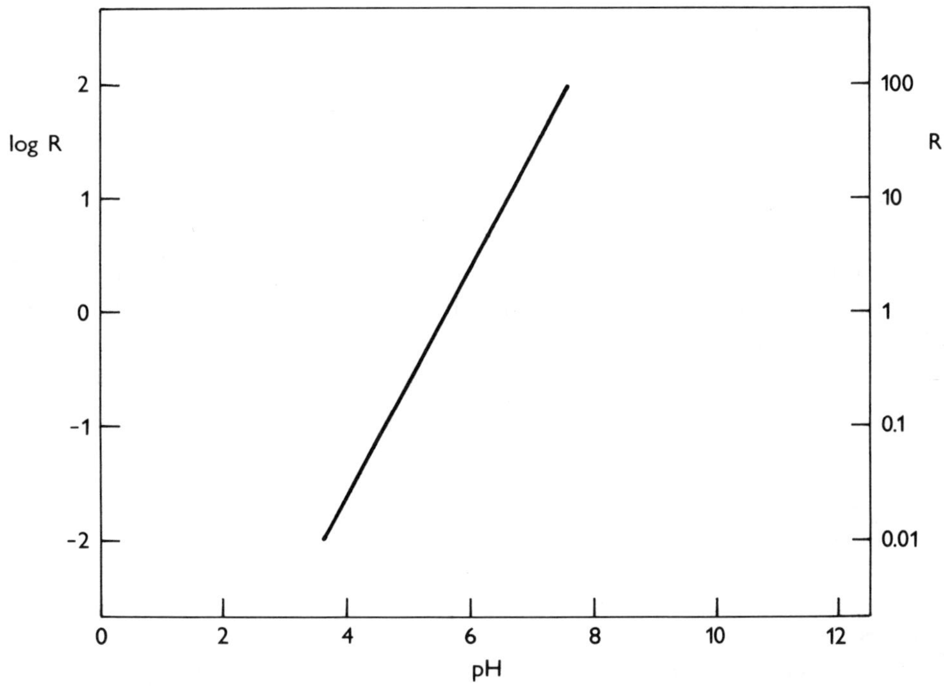

FIGURE 19. pH dependence of the ratio R of dissociated and undissociated forms of the indicator dye described in Figure 18.

The Hantzsch theory postulates the existence of two tautomeric forms (Ind_e and Ind_k) of the indicator, each exhibiting a different color. Both forms are (or at least one of them is) a weak acid (base). The situation is described by the three following chemical equations (Figures 20 and 21):

$$Ind_e \rightleftharpoons Ind_k \tag{119}$$

$$HInd_e + H_2O \rightleftharpoons Ind_e^- + H_3O^+ \tag{120}$$

$$HInd_k + H_2O \rightleftharpoons Ind_k^- + H_3O^+ \tag{121}$$

with corresponding equilibrium constants

$$K'_T = c_{Ind_k}/c_{Ind_e} \tag{122}$$

$$K'_1 = c_{Ind_e^-}c_{H3O+}/c_{HInd_e} \tag{123}$$

$$K'_2 = c_{Ind_k^-}c_{H3O+}/c_{HInd_k} \tag{124}$$

Based on the assumption that both the dissociated and the undissociated forms of the dye have the same absorption properties (the same color), we obtain (Figure 22)

$$R = \frac{c_{HInd_e} + c_{Ind_e^-}}{c_{HInd_k} + c_{Ind_k^-}} = \frac{(c_{H3O+} + K'_1)K'_2}{(c_{H3O+} + K'_2)K'_1K_T} \tag{125}$$

yielding

$$pH = -\log \frac{K'_1K'_2(1 - K'_TR)}{K'_1K'_TR - K'_2} \tag{126}$$

where R is again the value derived from absorption measurements as described in the previous section.

Use of a dye as a pH indicator requires one color form to dominate at acidic pH values and another at basic pH values (Figure 20). (The ratios c_{HInd_e}/c_{HInd_k} and $c_{Ind_e^-}/c_{Ind_k^-}$ are independent of pH as follows from Equations 122 to 124). For the substance to act as a pH indicator, the following conditions must be fulfilled (cf. Figures 20 to 22): either

$$K'_1 > 1, \quad K'_TK'_1/K'_2 < 1, \quad K'_1 \ll K'_2 \tag{127}$$

or, analogously,

$$K'_T < 1, \quad K'_TK'_1/K'_2 > 1, \quad K'_1 \gg K'_2 \tag{128}$$

H. The Isosbestic Point

From the spectroscopic point of view, an equilibration of two differently colored forms occurs according to both Ostwald and Hantzsch. Therefore, a wavelength must exist at which both forms have the same value of absorptivity and are spectroscopically indistinguishable. Absorption at this wavelength is pH independent. All absorption spectra (if pH is varied as a parameter) intersect at this wavelength, the point of intersection being called the isosbestic

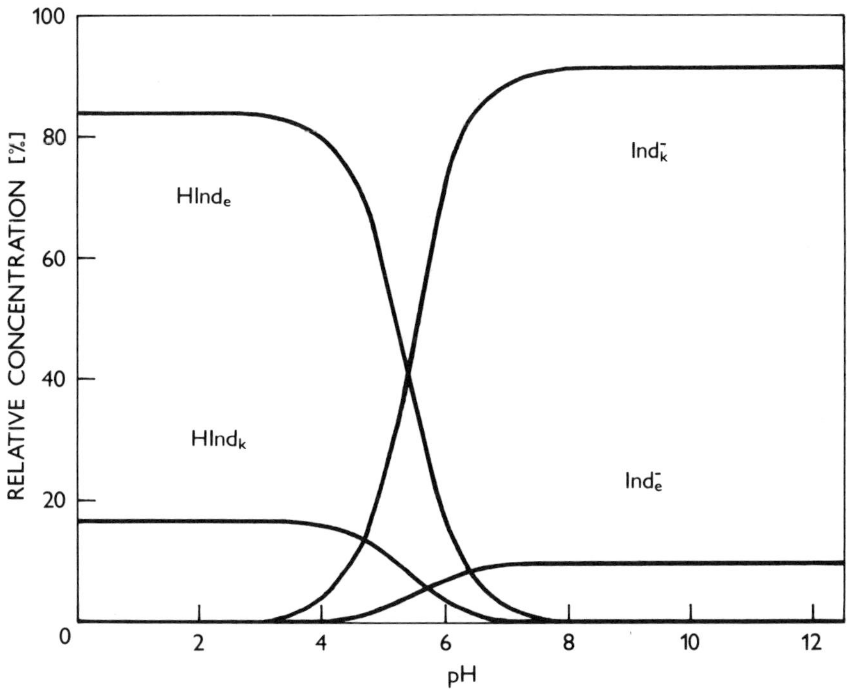

FIGURE 20. The relative concentrations of tautomeric forms of an acid and its conjugate base according to Hantzsch's theory (pK_1 = 6.7; pK_2 = 4.7; K_T = 10).

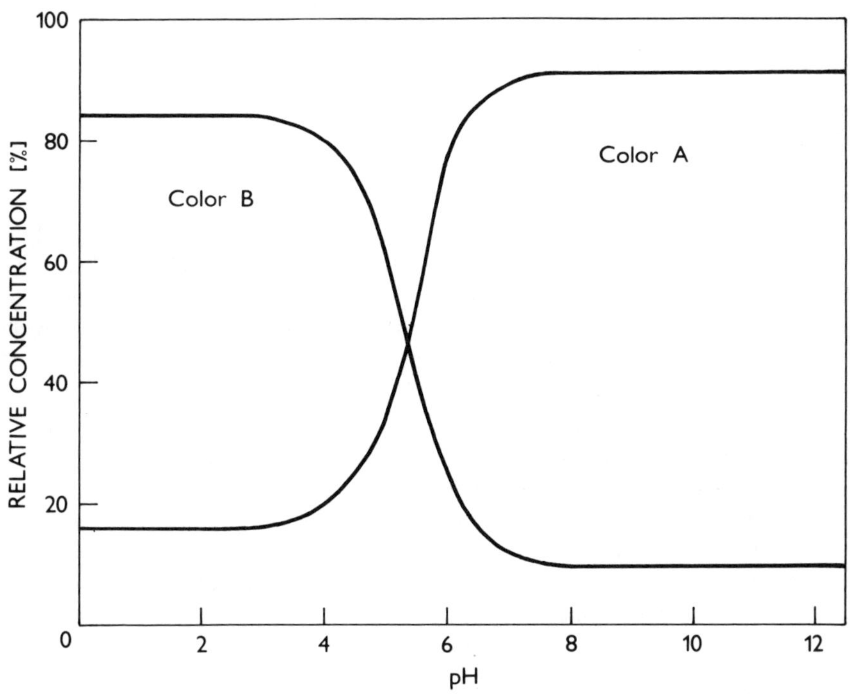

FIGURE 21. pH dependence of the absorptivity of the dye described in Figure 20, under the condition that the absorption peaks of the two colored forms do not overlap.

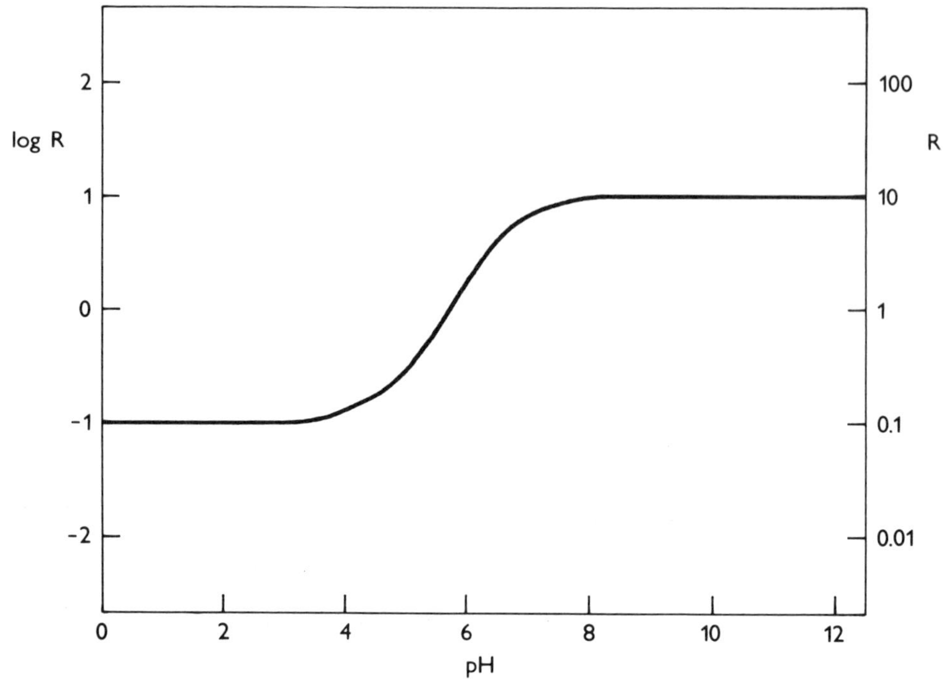

FIGURE 22. pH dependence of the ratio *R* of the two colored forms of the dye from Figure 20.

point. Its existence demonstrates that there are two equilibrating color forms; its absence suggests that there are more than two equilibrating forms.

II. ABSORPTION pH INDICATORS

A. Introduction

There are myriads of dyes exhibiting a significant pH-dependent change in color, and many of them are suitable as pH indicators (see Table 5 for a selection). However, the absorption measurements are relatively rare, and the number of dyes systematically used in intracellular pH measurements is small. The older works were revised by Caldwell,[2] recently by Roos and Boron,[3] and by Heiple and Taylor.[4] In recent experiments only fluorescein and carboxyfluorescein, neutral red, phenol red, bromothymol blue, bromocresol purple, and perhaps arsenazo III (which is both Ca^{2+} and pH sensitive) were repeatedly used.

B. Fluorescein and Carboxyfluorescein

The fluorochromasia phenomenon[5] (Chapter 4D.II.B.) was used for intracellular pH measurement by absorption 3 years earlier[6] than by fluorescence;[7,8] but since that time, fluorescence measurement is preferred to absorption.[1,9] Consequently, the data on fluorescein and its derivatives are presented in the chapter on fluorescence. Nevertheless, the introductory paper,[7] comparing the absorption and fluorescence approaches using fluorescein and carboxyfluorescein, represents a nice and interesting confrontation of the advantages of both methods.

C. Phenol Red

Phenol red (phenolsulfonphthalein) shows a pH color change between yellow and red in the pH range between 6.4 and 8.2. The corresponding absorption maxima of each form are

Table 5
COLOR RESPONSE OF VARIOUS pH INDICATORS

Indicator	pH range	Color change
Picric acid	0.1—1.3	Colorless—yellow
Brilliant green	0.1—2.6	Yellow—green
Crystal violet	0.1—2.0	Green—blue
Methyl violet	1.0—1.5	Yellow—violet
Cresol red	0.2—1.8	Red—yellow
Safranine T	0.3—1.0	Blue—red
Fuchsin	1.0—3.1	Yellow—magenta
Metanil yellow	1.2—2.3	Violet red—yellow
m-Cresol purple	1.2—2.8	Red—yellow
Thymol blue	1.2—2.8	Red—yellow
p-Xylenol blue	1.2—2.8	Red—yellow
Tropaeolin OO	1.2—3.2	Violet red—orange yellow
Quinaldine red	1.4—3.2	Colorless—pink
Methyl violet	1.5—3.2	Blue—violet
2,6-Dinitrophenol	1.7—4.4	Colorless—yellow
Benzyl orange	1.9—3.3	Red—yellow
2,4-Dinitrophenol	2.0—4.7	Colorless—yellow
Benzopurpurin 4B	2.3—4.4	Violet blue—red
Dimethyl yellow	2.9—4.0	Red—yellow
Congo Red	3.0—5.2	Blue—red
Bromophenol blue	3.0—4.6	Yellow—violet blue
Bromochlorophenol blue	3.0—4.6	Yellow—violet
Methyl orange	3.0—4.4	Red—orange yellow
Methyl orange	3.5—5.7	Violet pink—brownish yellow
α-Naphthyl red hydrochloride	3.7—5.0	Purple—orange yellow
Bromocresol green	3.8—5.4	Yellow—blue
2,5-Dinitrophenol	4.0—5.8	Colorless—yellow
Methyl red	4.4—6.2	Red—yellow
Ethyl red	4.4—6.2	Red—yellow
Chlorophenol red	4.6—7.0	Yellow—violet red
Carminic acid	4.8—6.2	Yellow—violet red
Hematoxylin	5.0—6.0	Yellow—violet
Alizarin red S	5.0—6.6	Yellow—violet red
2-Nitrophenol	5.0—7.0	Colorless—yellow
Litmus	5.0—8.0	Red—violet brown
Bromocresol purple	5.2—6.8	Yellow—violet
Bromophenol red	5.2—6.8	Yellow—purple
4-Nitrophenol	5.6—7.6	Colorless—yellow
Alizarin	5.8—7.2	Yellow—violet red
Bromothymol blue	6.0—7.5	Yellow—blue
Bromoxylenol blue	6.0—7.6	Yellow—blue
Brasiline	6.0—7.7	Greenish yellow—deep violet
Curcumin	6.0—8.0	Yellow—brown red
Phenol red	6.4—8.2	Yellow—red
3-Nitrophenol	6.6—8.6	Colorless—yellow
Neutral red	6.8—8.0	Red—yellow
Cresol red	7.0—8.8	Yellow—violet red
m-Cresol purple	7.4—9.0	Yellow—violet
Brilliant yellow	7.4—8.6	Yellow—brown red
Orange I	7.6—8.9	Yellow—pink
α-Naphtholphthalein	7.8—9.0	Bright yellow—blue
Thymol blue	8.0—9.6	Yellow—blue
p-Xylenol blue	8.0—9.6	Yellow—blue
o-Cresolphthalein	8.2—9.8	Colorless—violet red
Phenolphthalein	8.2—10.0	Colorless—purple
α-Naphtholbenzein	8.8—11.0	Colorless—green blue

Table 5 (continued)
COLOR RESPONSE OF VARIOUS pH INDICATORS

Indicator	pH range	Color change
Thymol violet	9.0—13.0	Yellow—violet
Thymolphthalein	9.3—10.5	Colorless—blue
Water blue	9.4—14.0	Blue—(red)—colorless
Alizarin yellow 2G	10.0—12.0	Bright yellow—orange yellow
Alizarin yellow R	10.0—12.0	Bright yellow—orange red
β-Naphthol violet	10.0—12.0	Orange yellow—violet
Nil blue A	10.2—13.0	Blue—violet red
Nitramin	10.8—12.8	Colorless—brown red
Tropaeolin O	11.1—12.7	Yellow—brown red
Alizarin	11.0—13.0	Pink—violet
Orange II	11.0—13.0	Yellow—red
Indigocarmine	11.6—14.0	Blue—bright yellow
Orange G	11.6—14.0	Yellow—pink
Thiazole yellow G	12.0—13.0	Yellow—brown yellow
Acid fuchsin	12.0—14.0	Magenta—colorless

360 and 557 nm. This was recently used by Lisman and Strong,[10] Baylor et al.,[11] Gylfe et al.,[12] Heiple and Taylor,[4] Ahmed and Connor,[13] and Bolsover et al.[14] Its high extinction makes it suitable for microspectrophotometry.

D. Neutral Red

Neutral red (3-amino-7-dimethylamino-2-methyl-phenazine) exhibits a red-yellow color change between pH 6.8 and 8.0. Its pH-indicating properties were studied by Pick and Avron[15] and lately used by MacDonald and Jöbsis,[16] Snow et al.,[17] Ausländer and Junge,[18] and Bashford.[19]

E. Bromocresol Purple

Bromocresol purple (5,5′-dibromo-*o*-cresolsulfonaphthalein) shows a yellow-violet color change between pH 5.2 and 6.8 with corresponding absorption maxima at 585 and 379 nm. It has been used for intracellular pH measurement several times, e.g., Ahmed and Connor,[13] MacDonald and Jöbsis,[16] MacDonald et al.,[20] Hersey,[21] and Hersey and High.[22]

F. Bromothymol Blue

Bromothymol blue (3′,3″-dibromothymolsulfonaphthalein) shows a yellow-blue color transition between pH 6.0 and 7.5 with corresponding maximal absorption wavelengths of 392 and 615 nm. Lately, it has been used by Hersey,[23,24] Piontek and Herbst,[25,26] and Carter.[27]

G. Arsenazo III

Recently, this indicator has been used several times,[11,13] but as its absorbance depends on both calcium (and magnesium) and proton concentration, the interpretation of data is not unambiguous.[11-13,28,29]

III. EXPERIMENTAL ASPECTS

A. Spectrophotometer

The basic arrangement for optical absorption measurements is extremely simple. The light emitted from a light source passes through the sample and is detected by a light detector (Figure 23).

Modern measuring instruments employ more than one light source (each interchangeable

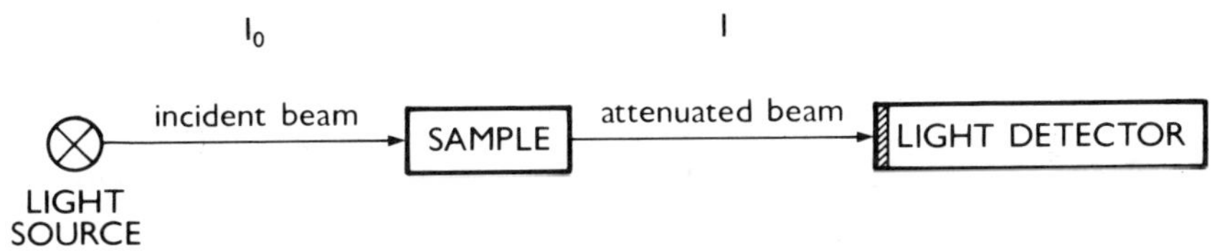

FIGURE 23. The basic principle of light absorption measurements.

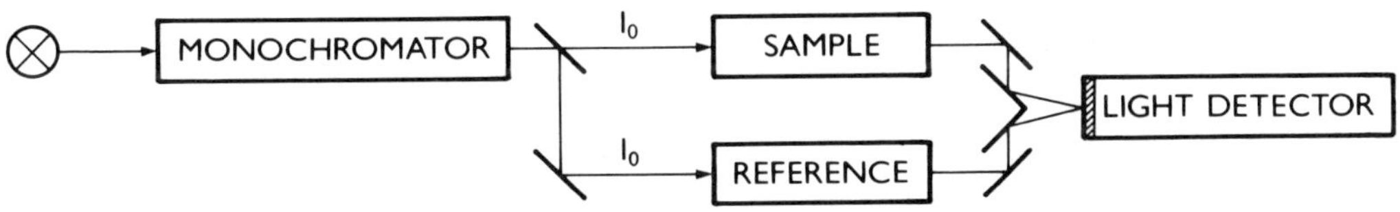

FIGURE 24. Schematic drawing of a double-beam spectrophotometer.

for a specific wavelength range), a prism or a grating monochromator, and/or optical filters. The light beam is mostly split into two equivalent beams, one of which passes through the sample, the other serving as a reference (Figure 24). These so-called double-beam spectrophotometers (not to be confused with double-wavelength spectrometers to be discussed below) are preferred to single-beam ones as they make it possible to compare the absorption of the sample with a ''blank'', represented by an empty cuvette, pure solvent, or cells without dye. The absorption spectrum is displayed on a plotter or a TV monitor in the desired form (absorbance, transmittance, etc. vs. wavelength or frequency), mostly with numerical values of the individual peaks already assigned.

There is a special modification of the absorption spectrophotometer making it possible to measure absorption at two selected wavelengths simultaneously. This is the so-called double-wavelength spectrophotometer, which is especially suited for intracellular pH measurement.

A number of firms produce reliable and reasonably priced double-beam spectrophotometers (e.g., Shimadzu, Philips, Perkin-Elmer, Beckman, Jobin-Yvon, LKB, Milton Roy, Varian, Spex, and Sigma, to give a selection). Because of the similarity between absorption and fluorescence principles, an apparatus is made by Spex (the Fluorolog®) capable of both absorption and fluorescence measurements of a sample in sequence.

The basic arrangement shown in Figure 24 can be largely modified; e.g., the sample may be put between microscopic glasses and investigated under the optical microscope equipped with a photographic or a TV camera as a photodetector. Such modifications could give the estimates of intracellular pH of individual cells or tissue sections, or even intracellular pH topography of cells (cf. section on microfluorometry).

The diode array spectrophotometer, which makes it possible to detect light intensity simultaneously at 1024 different wavelengths and, thus, to record the whole absorption spectrum at once, was recently introduced into intracellular pH estimations.[9,30]

B. Techniques of Measurement

Double-beam instruments are preferred to single-beam ones, the reference sample consisting of cells without dye. The cell suspension density and the concentration of the dye should be chosen so that the samples appear almost clear to the naked eye (optical densities below 0.1). The use of turbid samples should be avoided, if possible. In any case, the scattering of light (diffractions, internal reflections on myriads of optical boundaries) represents a serious problem which may be difficult to counter. The application of barrier filters and of a double monochromator may exclude the contribution of false-wavelength light and

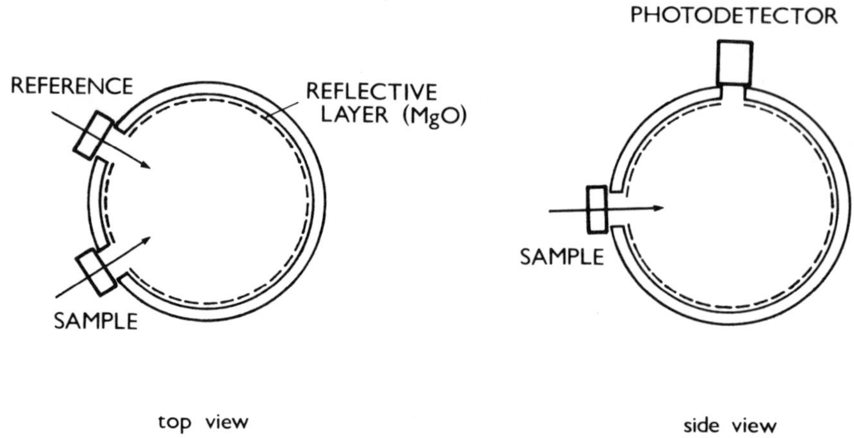

FIGURE 25. The integrating sphere used for measurement of diffusive absorption in a double-beam technique. Because of the nearly 100% reflectivity of the internal wall surface, virtually all light that has passed through the sample is registered by the photodetector.

make the measurement of slightly turbid sample feasible. The use of an integrating sphere (occasionally called the Ulbricht sphere) and the measurement of diffusive absorption can also help significantly (Figure 25).

C. Sensitivity, Temporal, and Spatial Resolution

Sensitivity of the absorption techniques can be calculated by the Lambert-Beer Law, provided that absorbance values around 0.0001 to 0.001 are needed for a reasonable measurement. For high-absorbing dyes this means a 10 n*M* concentration, for low-absorbing ones ($\epsilon = 10^3$) about 1 μ*M*. For microphotometric measurements, the concentration of the dye in cells should be 100 or more times higher. The main reason is that the length of the light path in the microscope is much shorter, about 10 μm compared to 1 cm in the cuvette (i.e., 10^3 times).

The temporal resolution of absorption is virtually unlimited. The process of absorption is almost instantanous (1 fs) and certainly several orders of magnitude faster than the pH response of any pH-indicating dye could be. Therefore, the equilibration of the dye with the surrounding medium is the limiting step for the temporal resolution (estimated at 10 μs). However, the response of the electronic part of a routine spectrophotometer is much slower: for a classical double-wavelength apparatus it is around 1 ms; for a diode array spectrophotometer about 100 ms. If faster pH changes are to be followed, the detection part must be rebuilt. Then the temporal resolution of the apparatus is limited by the response of the photomultiplier (typically 1 ns).

The spatial resolution is limited by the wave properties of light. The resolution limit of a good optical microscope is $\sim 0.4\lambda$ (0.15 to 0.2 μm).

D. How to Prepare a Biological Sample

The requirements and limitations imposed by absorption measurements are in many ways similar to those of fluorescence measurements. Therefore, many ways of treating a biological sample for obtaining reliable results are similar to fluorescence. Not only the technical aspects, but, in some cases, even the dye used is the same (e.g., fluorescein). The major difference is that the dye concentrations are generally higher for absorption measurements. Cells requiring attachment to a substratum can be grown on transparent plastic Leighton tube slides which directly fit the cuvette of the spectrophotometer[1] (cf. Chapter 4D, Figure 46).

E. Interpretation of Results

Because of the simplicity of the technique, no serious problems should occur here. The absorption spectrum of an intracellular dye should resemble the absorption of the dye in solution. The additivity of absorption is expressed already in the Lambert-Beer Law, and, therefore, the background signal can be subtracted whenever required.

F. Problems of pH Calibration

There are almost no pH-calibration problems specific to the absorption measurement. The standard calibration procedure can be performed as described in Chapter 4A. Possible pitfalls are (1) the aggregation of the dye if a high concentration of the dye is used and (2) the change of pH measured by the very presence of the dye. If the "right" concentration is used, no difficulties should occur. If these problems are avoided, the accuracy is determined by the stability of p*K*. Equation 78 shows that any shift in p*K* results in the same shift in the pH value estimated. Many aspects of pH calibration presented in the chapter about fluorescence techniques and microspectrofluorometry apply to absorption measurements.

In microphotography, there are minor differences as to the preference of films: medium-sensitive films and dye films are preferred to silver films. The stability of the densitometric curve of the film or the stability of the sensitivity of TV camera or photomultiplier limits the sensitivity of microphotometric measurements. The densitometric comparison of a picture pair (similar to the computerized image analysis of two pictures obtained by selective excitation in the fluorescence intracellular-pH topography procedure, cf. Chapter 4D.V.) is equivalent to double-wavelength single-beam absorption spectroscopy. Perhaps, an introduction of a third picture with no cells or with unstained cells could represent a blank, and the signal could be used as the second beam in double-beam spectrophotometers.

G. A Typical Experimental Result

An example of absorption spectra of a biological sample employed for intracellular pH measurement is shown in Figure 26 (top). The background was compensated by using cells of the same concentration but unstained, and placing them in the reference cuvette. The pH value of 5.8 was derived by computation of the difference in absorbances at $\lambda_1 = 435$ nm and $\lambda_2 = 490$ nm, $\Delta A = 0.9 - 0.4 = 0.5$ and by comparison of ΔA with the values in the calibration curve (Figure 26, bottom). The following steps were taken:

1. The dye was chosen according to the properties described in Section C.2 (fluorescein).
2. The non-interference of the dye with physiological processes was checked (200 μM concentration used).
3. The dye was formed *in situ* from fluorescein diacetate ("trapped indicator").
4. pH calibration was done *in situ* using ^{14}C-propionic acid or bromophenol blue distribution.
5. The leak of the dye was minimized by rapid centrifugation prior to measurement.

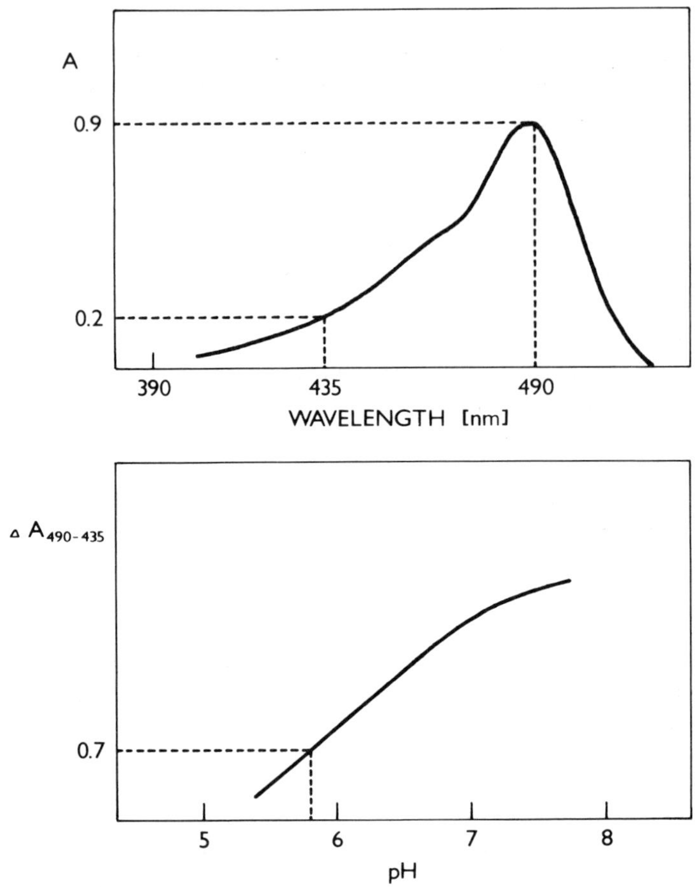

FIGURE 26. A typical result of absorption measurement of intracellular pH (above) with the corresponding pH standard curve (below). For details see the text.

REFERENCES

1. **Kaplan, A. E. and Bunow, M. R.,** Spectrophotometric determination of intracellular pH with cultured rat liver epithelial cells, *J. Histochem. Cytochem.*, 34, 749, 1986.
2. **Caldwell, P. C.,** Intracellular pH, *Int. Rev. Cytol.*, 5, 229, 1956.
3. **Roos, A. and Boron, W. F.,** Intracellular pH, *Physiol. Rev.*, 61, 296, 1981.
4. **Heiple, J. M. and Taylor, D. L.,** An optical technique for measurement of intracellular pH in single living cells, in *Intracellular pH: Its Measurement, Regulation and Utilization in Cellular Functions,* Nuccitelli, R. and Deamer, D. W., Eds., Alan R. Liss, New York, 1982, 21.
5. **Rotman, D. and Pappenmaster, B. W.,** Membrane properties of living mammalian cells as studied by enzymatic hydrolysis of fluorogenic esters, *Proc. Natl. Acad. Sci. U.S.A.*, 55, 134, 1966.
6. **Thomas, J. A., Buschbaum, R. N., Zimniak, A., and Racker, E.,** Intracellular pH measurements in Ehrlich ascites tumor cells utilizing spectroscopic probes generated in situ, *Biochemistry,* 18, 2210, 1979.
7. **Thomas, J. A., Kolbeck, P. C., and Langworthy, T. A.,** Spectrophotometric determination of cytoplasmic and mitochondrial pH transitions using trapped pH indicators, in *Intracellular pH: Its Measurement, Regulation and Utilization in Cellular Functions,* Nuccitelli, R. and Deamer, D. W., Eds., Alan R. Liss, New York, 1982, 105.
8. **Slavík, J.,** Intracellular pH of yeast cells measured with fluorescent probes, *FEBS Lett.*, 140, 22, 1982.

9. **Chaillet, J. R. and Boron, W. F.,** Intracellular calibration of a pH sensitive dye in isolated, perfused salamander proximal tubules, *J. Gen. Physiol.,* 86, 765, 1985.
10. **Lisman, J. E. and Strong, J. A.,** The initiation of excitation and light adaptation in Limulus ventral photoreceptors, *J. Gen. Physiol.,* 73, 219, 1979.
11. **Baylor, S. M., Hollingworth, S., and Pape, P.,** Myoplasmic pH transients monitored with indicator dyes in frog skeletal muscle fibers, *Biophys. J.,* 51, 549a, 1987.
12. **Gylfe, E., Berggren, P. O., and Rorsman, P.,** Measurements of cytoplasmic pH in insulin-releasing cells with intracellularly trapped phenol red, *Biomed. Res.,* 7, 139, 1986.
13. **Ahmed, Z. and Connor, J. A.,** Intracellular pH changes induced by calcium influx during electrical activity in molluscan neurons, *J. Gen. Physiol.,* 75, 403, 1980.
14. **Bolsover, S. R., Brown, J. E., and Goldsmith, T. H.,** Intracellular pH of Limulus ventral photoreceptor cells — measurement with phenol red, in *Optical Methods in Cell Physiology,* Deweer, P. and Salzberg, B. M., Eds., John Wiley & Sons, New York, 1986.
15. **Pick, U. and Avron, M.,** Neutral red response as a measure of the pH gradient across chloroplast membranes in the light, *FEBS Lett.,* 65, 348, 1976.
16. **MacDonald, V. W. and Jöbsis, F. F.,** Spectrophotometric studies on the pH of frog skeletal muscle, *J. Gen. Physiol.,* 68, 179, 1976.
17. **Snow, T. R., Saive, J. J., and Hemstreet, T. M.,** A study of the temporal relation between intracellular pH and contractile performance in toad ventricular strips during hypercapnic acidosis, *J. Mol. Cell. Cardiol.,* 14, 1, 1982.
18. **Ausländer, W. and Junge, W.,** Neutral red, a rapid indicator for pH changes in the inner phase of thylakoids, *FEBS Lett.,* 59, 310, 1975.
19. **Bashford, C. L.,** Spectrophotometry and fluorimetry of cellular compartments, in *Spectrophotometry and Spectrofluorimetry — A Practical Approach,* Harris, D. A. and Bashford, C. L., Eds., IRL Press, Oxford, 1987, 115.
20. **MacDonald, V. W., Keizer, J. H., and Jöbsis, F. F.,** Spectrophotometric measurements of metabolically induced pH changes in frog skeletal muscle, *Arch. Biochem. Biophys.,* 184, 423, 1977.
21. **Hersey, S. J.,** The energetic coupling of acid excretion in gastric mucosa, *Philos. Trans. R. Soc. London Ser. B,* 262, 261, 1971.
22. **Hersey, S. J. and High, W. L.,** On the mechanism of acid secretory inhibition by acetazolamide, *Biochim. Biophys. Acta,* 233, 604, 1971.
23. **Hersey, S. J.,** Intracellular pH and gastric acid secretion, *Acta Physiol. Scand. Spec. Suppl.,* 243, 1978.
24. **Hersey, S. J.,** Intracellular pH measurements in gastric mucosa, *Am. J. Physiol.,* 237, E82, 1979.
25. **Piontek, P. and Herbst, M.,** Probleme der photometrischen Messung mit Farbindikatoren an Skeletmuskeln, *Pflügers Arch. Gesamte Physiol. Menschen Tiere,* 328, 356, 1971.
26. **Herbst, M. and Piontek, P.,** Über den Verlauf des intrazellulären pH — Wertes des Skeletmuskels während der Kontraktion, *Pflügers Arch. Gesamte Physiol. Menschen Tiere,* 335, 213, 1972.
27. **Carter, N. W.,** Intracellular pH, *Kidney Int.,* 1, 341, 1972.
28. **Hockberger, P. and Connor, J. A.,** Intracellular calcium measurements with arsenazo III during cyclic AMP injections into molluscan neurons, *Science,* 219, 869, 1983.
29. **Ahmed, Z. and Connor, J. A.,** Measurement of calcium influx under voltage clamp in molluscan neurones using the metallochromic dye Arsenazo III, *J. Physiol.,* 286, 61, 1979.
30. **Konishi, T., Murakami, N., Hatano, Y., and Nakazato, K.,** Simultaneous determination of membrane potential and pH gradient by photodiode array spectroscopy, *Biochem. Biophys. Acta,* 862, 278, 1986.

Chapter 4D

ASSAYS OF INTRACELLULAR pH USING CHEMICAL PROBES: FLUORESCENCE SPECTROSCOPY

Jan Slavík

TABLE OF CONTENTS

I. THEORETICAL ASPECTS

A. Introduction

Like absorption spectroscopy, fluorescence spectroscopy has been a traditional analytical technique. Its popularity was high in the 1920s and 1930s, then waned somewhat. It reappeared again, together with fluorescent probes, some 20 years ago.

The main advantage of fluorescence lies in its high sensitivity, particularly in comparison to absorption and NMR techniques, and this has brought fluorescence techniques into the foreground of interest in biochemistry, biology, and medicine. Other strong points of fluorescence are similar to those of absorption. In the first place, it is possible to see and distinguish spectroscopically small areas of the sample, within the spatial resolution limit of the optical microscope (150 to 200 nm). The temporal resolution is slower than that of the absorption techniques (10 ns vs. 1 fs), although, in practice, this hardly makes any difference. The disadvantages may be a slightly more complicated theory and instrumentation than that of the absorption technique. However, as all modern experimental techniques become more and more complicated each year, this disadvantage becomes less important.

Nevertheless, the basic principles are easy to grasp as they are very similar to those of absorption techniques. A concise introduction to the fluorescence of solutions can be found, e.g., in the books by Lakowitz[1] or Parker.[2]

B. Fluorescent Compounds

Only a few of those compounds that absorb ultraviolet or visible light exhibit a sufficiently strong fluorescence (though this number can be increased a low temperatures). To predict on the basis of the chemical formula whether a particular compound will fluoresce or not is even more difficult than in the case of absorption. For instance, fluorescein is a strongly fluorescent and colored dye, while its derivative, fluorescein diacetate, is non-fluorescent and colorless in solution.

Like absorption, fluorescence is attributable to specific chemical groups called chromophores or fluorochromes. In practice, these are arbitrarily divided into fluorescent probes and fluorescent labels (markers).

Fluorescent probes (e.g., 1-anilinonaphthalene-8-sulfonate [ANS] or 1,6-diphenyl-1,3,5-hexatriene [DPH]) are molecules (or groups) with fluorescence highly sensitive to their environment.[3] They can be employed as tiny molecular reporters — ''spies'' which, encoded in the fluorescence signal, pass on information about their molecular neighborhood such as polarity, viscosity, presence, and movement of other chromophores and, of course, about pH. Fluorescent probes are not thought to be covalently bound to the object (or area) of interest.

Fluorescent labels, or markers (e.g., fluorescein or rhodamine), are fluorochromes which exhibit intense fluorescence virtually independent of their molecular neighborhood. They are covalently bound to the selected molecule (e.g., immunoglobulin), the behavior of which is to be followed.

Auspiciously, the fluorescence of one of the most famous fluorescent labels, fluorescein, is sensitive to pH, with p*K* just in the middle of the physiological pH range (p*K* $\simeq$6.8). In principle, all molecules marked with fluorescein can, thus, be employed for intracellular-pH topography. (Imagine the wide range of monoclonal fluorescein-labeled antibodies!)

C. The Phenomenon of Fluorescence

The energy absorbed by a molecule due to light absorption (but also, because of chemical reaction and mechanical movement) is often transiently stored in the form of energy of electrons. The return of an electron from an energetically higher (excited) state to the initial, energetically lower (ground) state may be accompanied by the release of a photon (Figure 27). The energy of this photon is lower or equal to the energy of the absorbed photon (Stokes' relation)

$$h\nu_{abs} \geqslant h\nu_{flu} \tag{129a}$$

or expressed in wavelengths

$$\lambda_{abs} \leqslant \lambda_{flu} \tag{129b}$$

The properties of fluorescence photons are almost independent of the way the electron was brought into the excited state, since only transitions from the lowest excited state to the ground state ($S_1 \rightarrow S_0$) may be accompanied by fluorescence (Kasha's rule). Therefore, fluorescence parameters reflect the properties of both the ground state S_0 and the lowest

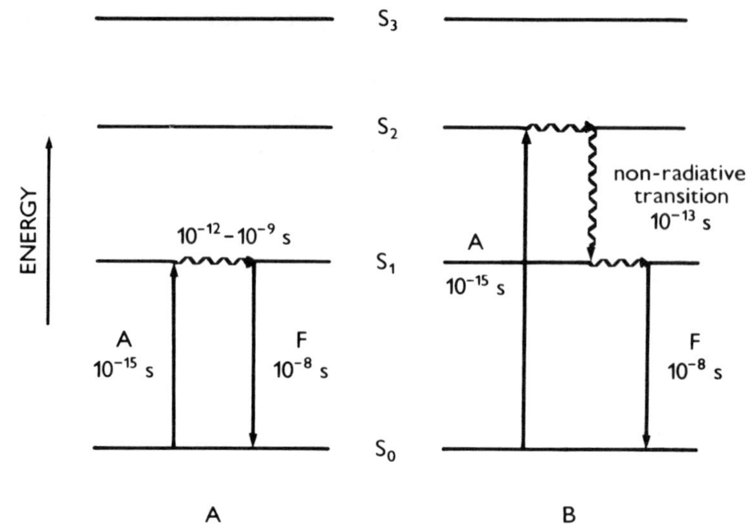

FIGURE 27. Occupancy of electron energy levels and the process of absorption and emission of light. Part A shows excitation to the lowest (S_1) excited state; part B, to the next higher (S_2) excited state.

excited state S_1, which may differ slightly. (The higher chemical reactivity in the excited state and the pK difference is usually significant).

D. What is Measured?

1. Intensity of fluorescence — the number of emitted quanta (rarely their total energy) in the given wavelength range per unit time
2. Emission spectrum (''fluorescence spectrum'') — the dependence of the intensity of fluorescence upon the fluorescence wavelength (at a given excitation wavelength)
3. Excitation spectrum — the dependence of the intensity of fluorescence upon the excitation wavelength (at a fixed fluorescence wavelength)
4. Quantum yield — the number of quanta emitted per number of quanta absorbed (''efficiency'' of fluorescence) with values ranging between 0 and 1
5. Lifetime — the mean time (statistically) spent by an average chromophore in the excited state (typically 10 ns)
6. Polarization of fluorescence — the degree of polarization (or anisotropy) of emitted light with values ranging between 0 and a limiting value (typically 0.4)
7. Time-resolved fluorescence (TRF) — in addition to the usual ''steady-state'' fluorescence, the temporal changes of excitation or emission spectra, or polarization; followed on a time scale comparable to the lifetime of the chromophore

E. Basic Relations between Fluorescence Parameters

1. Introduction

There is a large number of relations combining various fluorescence parameters. These relations are useful in two ways. First, they may be employed as a check on what has actually been measured. Second, a careful combination of different experimental data can yield a large amount of information on the behavior of molecules, both in their ground and excited states.

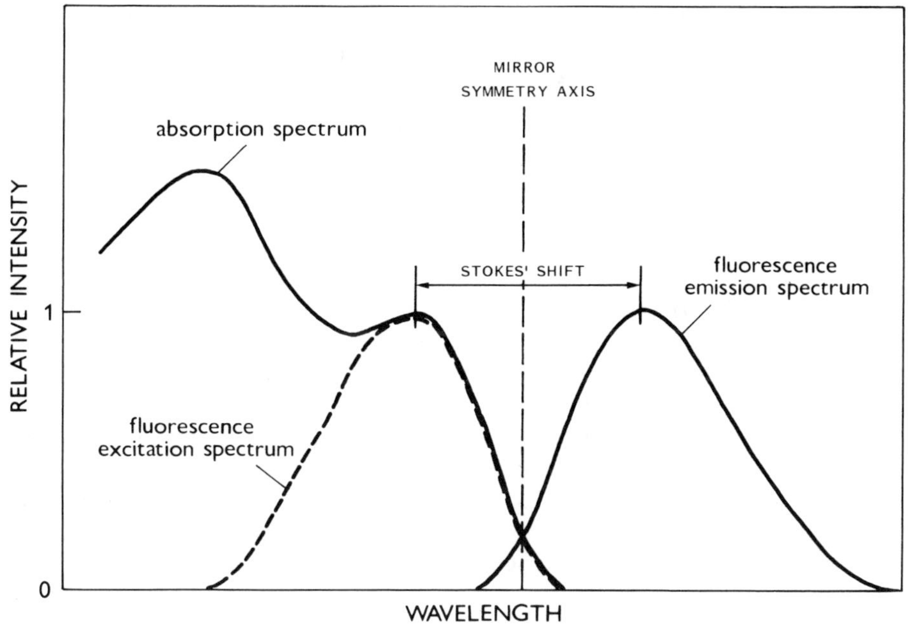

FIGURE 28. Relation between the normalized fluorescence excitation and emission spectra and the absorption spectrum. The lower peak of the absorption spectrum corresponds to S_1, the higher one to S_2 of Figure 27.

2. Shape of the Spectrum

The most important point is that the shape of excitation, absorption, and emission spectra must be similar if they all arise from a single compound. If not, probably some kind of impurity is present (very low concentrations can thus be detected). There must be a mirror symmetry between the absorption and emission spectra, and the excitation spectrum should resemble the shape of the absorption spectrum (Figure 28).

In fact, this applies only to the part of the absorption spectrum corresponding to the first absorption band, and, unlike with NMR and absorption measurements, the shape of the fluorescence spectrum depends on the technical properties of the apparatus on which the spectrum is recorded. Every instrument has a different spectral sensitivity, but only a few use a correction procedure. The result is called corrected excitation or emission spectrum.

3. Fundamental Equation for Fluorescence Intensity

The basic equation of fluorescence is similar to the Lambert-Beer Law as it relates the intensity of fluorescence I_F to that of incident light I_O, concentration of molecules c, their molar absorptivity ϵ, and the quantum yield γ

$$I_F = kI_0 c\epsilon\gamma \tag{130}$$

where kI is the instrument constant. (Note that the equation is only accurate for small light intensities.) According to this equation, an increase in fluorescence intensity may arise from three different causes:

1. Higher absorption due to a more convenient choice of excitation parameters (another wavelength, rarely another polarization of light)
2. Trivial increase due to the increase of dye concentration
3. An increase in quantum yield (e.g., some fluorescence-quenching process is stopped)

The first case requires comparison with absorption or excitation spectra, the second case manifests itself in the absorption spectrum, and the third case is accompanied by an increase in the lifetime, which can be measured.

4. Lifetime and Quantum Yield

The lifetime τ and the quantum yield γ are related by

$$\gamma = \gamma_0 \tau \tag{131}$$

where γ_o is the (idealized) highest possible quantum yield of the compound.

The decrease in the lifetime (or quantum yield) is usually a consequence of an encounter of a fluorescent molecule with another partner. It may be the molecule of the same type (self-quenching) or another type. It involves formation of dimers, aggregates, and excimers; binding; transfer of excitation energy to another molecule (a donor chromophore is quenched but the acceptor molecule fluoresces); and many other processes. Some substances have a pronounced quenching effect even if present at low concentrations (the so-called quenching substances, such as ions Br^-, I^-, and Cl^-, but also dinitrophenol and CCCP). The above-mentioned processes are often diffusion-controlled and, thus, viscosity-dependent.

5. Time-Resolved Fluorescence

Time-resolved fluorescence allows a detailed inspection of these processes, as it makes it possible to follow, step by step, the behavior of an excited molecule; e.g., motional restrictions of bound molecules can be studied in this way based on the statistical nature of fluorescence decay. (Some molecules of an excited group decay sooner, some later, with the mean lifetime τ).

F. Theory of pH-Induced Fluorescence Change

1. Fundamental Equation

The basic principle of the pH dependence of fluorescence is the same as that of a color change of absorption pH indicators. As in absorption, the Hantzsch theory should probably be preferred to the Ostwald theory (see Chapter 4C.I.G.). Moreover, the similarity goes further as in Equation 130, determining the fluorescence intensity, is formally similar to the Lambert-Beer Law. Thus, equations formally similar to Equations 100 and 101 result:

$$I_F^{\lambda_1\lambda_3} = \epsilon_{\lambda_1}^{A}\gamma_{\lambda_3}^{A}c_A k I_O^{\lambda_1} + \epsilon_{\lambda_1}^{A}\gamma_{\lambda_3}^{B}c_B k I_O^{\lambda_1} \tag{132}$$

$$I_F^{\lambda_2\lambda_4} = \epsilon_{\lambda_2}^{A}\gamma_{\lambda_4}^{A}c_A k I_O^{\lambda_2} + \epsilon_{\lambda_2}^{B}\gamma_{\lambda_4}^{B}c_B k I_O^{\lambda_2} \tag{133}$$

where I_F is the intensity of fluorescence, I_O the intensity of exciting light, ϵ^A and ϵ^B are molar absorptivities, γ^A and γ^B the quantum yields, and c_A and c_B the concentrations of each form. The corresponding excitation (λ_1, λ_2) and emission (λ_3, λ_4) wavelengths are represented by subscripts; k characterizes the overall apparatus constant. In principle, one can choose any combination — both excitation and fluorescence wavelengths can be varied. However, in practice, either the pH dependence of excitation ($\lambda_1 \neq \lambda_2$, $\lambda_3 = \lambda_4$) or the pH dependence of emission ($\lambda_1 = \lambda_2$, $\lambda_3 \neq \lambda_4$) is used.

2. Double-Excitation Approach

In the first case, excitation is carried out at two different wavelengths, while fluorescence is measured at the same wavelength ($\lambda_3 = \lambda_4$).

Equations 132 and 133 then simplify to

$$I_F^{\lambda_1} = \epsilon_{\lambda_1}^A \gamma^A c_A k I_O^{\lambda_1} + \epsilon_{\lambda_1}^B \gamma^B c_B k I_O^{\lambda_1} \quad (134)$$

$$I_F^{\lambda_2} = \epsilon_{\lambda_2}^A \gamma^A c_A k I_O^{\lambda_2} + \epsilon_{\lambda_2}^B \gamma^B c_B k I_O^{\lambda_2} \quad (135)$$

In analogy to Equation 102, we obtain

$$I_F^{\lambda_1}/I_F^{\lambda_2} = (\epsilon_{\lambda_1}^A \gamma^A + \epsilon_{\lambda_1}^B \gamma^B R)/(\epsilon_{\lambda_2}^A \gamma^A + \epsilon_{\lambda_2}^B \gamma^B R) \quad (136)$$

and, in analogy to Equation 103,

$$\mathrm{pH} = \mathrm{p}K' + \log(-\epsilon_{\lambda_1}^A \gamma^A + \epsilon_{\lambda_1}^B \gamma^B I_F^{\lambda_1}/I_F^{\lambda_2})/(\epsilon_{\lambda_2}^A \gamma^A - \epsilon_{\lambda_2}^B \gamma^B I_F^{\lambda_1}/I_F^{\lambda_2}) \quad (137)$$

3. Double-Emission Approach

In the second case, excitation is done at one wavelength ($\lambda_1 = \lambda_2$) for both fluorescence wavelengths, but fluorescence is detected at two different wavelengths, λ_3 and λ_4.

Equations 132 and 133 then simplify to

$$I_F^{\lambda_3} = \epsilon^A \gamma_{\lambda_3}^A c_A k I_O + \epsilon^B \gamma_{\lambda_3}^B c_B k I_O \quad (138)$$

$$I_F^{\lambda_4} = \epsilon^A \gamma_{\lambda_4}^A c_A k I_O + \epsilon^B \gamma_{\lambda_4}^B c_B k I_O \quad (139)$$

In analogy to Equation 102 (for $\lambda_1 = \lambda_2$),

$$I_F^{\lambda_3}/I_F^{\lambda_4} = (\epsilon^A \gamma_{\lambda_3}^A + \epsilon^B \gamma_{\lambda_3}^B R)/(\epsilon^A \gamma_{\lambda_4}^A + \epsilon^B \gamma_{\lambda_4}^B R) \quad (140)$$

and Equation 103 transforms into

$$\mathrm{pH} = K' + \log(-\epsilon^A \gamma_{\lambda_3}^A + \epsilon^B \gamma_{\lambda_3}^B I_F^{\lambda_3}/I_F^{\lambda_4})(\epsilon^A \gamma_{\lambda_4}^A - \epsilon^B \gamma_{\lambda_4}^B I_F^{\lambda_3}/I_F^{\lambda_4}) \quad (141)$$

Equations 137 and 141 are identical to 103, but ϵ in the absorption equations must be replaced with $\epsilon\gamma$. Naturally, under the condition that

$$\epsilon_{\lambda_1}^A \gamma^A \ll \epsilon_{\lambda_1}^B \gamma^B \quad \text{and} \quad \epsilon_{\lambda_2}^A \gamma^A \ll \epsilon_{\lambda_2}^B \gamma^B \quad (142)$$

(the case of selective excitation), or

$$\epsilon^A \gamma_{\lambda_3}^A \ll \epsilon^B \gamma_{\lambda_3}^B \quad \text{and} \quad \epsilon^A \gamma_{\lambda_4}^A \ll \epsilon^B \gamma_{\lambda_4}^B \quad (143)$$

(the case of selective emission), the pH-induced fluorescence intensity changes are larger, and the determination of pH is easier (hence, in view of the background fluorescence, is more precise).

Naturally, the excitation wavelengths λ_1, λ_2 and emission wavelengths λ_3, λ_4 should be chosen to ensure optimally selective monitoring of the two dye forms and their clear distinction from a possible fluorescence background. It may take a lot of effort to find the optimal combination and to decide whether to distinguish the two forms of the dye on the basis of differences in the excitation or emission spectrum.

A special case, which is not rare in fluorescence, is that one form of the dye has a negligible quantum yield (almost no fluorescence is observed). In such a case, $\gamma_B = 0$, and Equations 133 and 134 simplify to

$$I_F^{\lambda_1\lambda_3} = \epsilon_{\lambda_1}^A \gamma_{\lambda_3}^A c_A k I_O^{\lambda_1} \tag{144}$$

$$I_F^{\lambda_2\lambda_4} = \epsilon_{\lambda_2}^A \gamma_{\lambda_4}^A c_A k I_O^{\lambda_2} \tag{145}$$

Evidently, it makes no sense to measure at two wavelengths as in both cases only information about c_A is obtained. On putting $\lambda_1 = \lambda_2$ and $\lambda_3 = \lambda_4$, Equations 144 and 145 become identical and, hence, a single-wavelength approach is warranted.

4. Single-Wavelength Approach

The single-wavelength approach to fluorescence pH measurement is quite often used in spite of the low reliability of its results. However, for pH indicators where only one form of the dye fluoresces, the single-wavelength approach is the only choice. In analogy to Equation 142, Equation 132 yields (for selected excitation and emission wavelengths)

$$I_F = \epsilon^A \gamma^A c_A k I_O + \epsilon^B \gamma^B c_B k I_O \tag{146}$$

and the measurement at the second wavelength is replaced by the assumption of constant total concentration of the dye c_T

$$c_T = c_A + c_B \tag{147}$$

Solving these two equations for c_A and c_B, we obtain

$$R = c_B/c_A = (\epsilon^A\gamma^A - \epsilon^B\gamma^B)/(I_F/c_T k I_O - \epsilon^B\gamma^B) - 1 \tag{148}$$

and, from pH $=$ pK' $+$ log R, the calibration curve follows:

$$\text{pH} = \text{p}K' + \log\left(\frac{\epsilon^A\gamma^A - \epsilon^B\gamma^B}{I_F/c_T k I_O - \epsilon^B\gamma^B} - 1\right) \tag{149}$$

In the special case of nonfluorescent form B, the condition $\gamma^B = 0$ further simplifies Equation 149 to

$$\text{pH} = \text{p}K' + \log(\epsilon^A\gamma^A k I_O/I_F - 1) \tag{150}$$

The single-wavelength approach in fluorescence has the same disadvantages and weaknesses as in absorption. Its use should be restricted to routine measurements only, and the results should be checked occasionally by the double-wavelength approach or another pH-measuring technique. However, if the conditions of constant total dye concentration and constant light path are fulfilled, the single-wavelength technique is invaluable for measuring fast pH changes.

G. The Isosbestic Point

Because of the similarity between the equation defining absorbance of several forms of a dye in solution (cf. Equation 99)

$$A = \sum_i \epsilon_i c_i l \tag{151}$$

and that of fluorescence intensity (cf. Equation 130)

$$I_F = I_O \sum_i \epsilon_i c_i \gamma_i k \tag{152}$$

the isosbestic point should appear in both excitation and emission fluorescence spectra under the same conditions as in absorption spectra.

II. FLUORESCENT pH INDICATORS

A. Introduction

There are an enormous number of pH-sensitive fluorescent dyes that could be employed as intracellular pH indicators. Many of them would probably do better than the fluorescent pH indicators presently used. However, as the characterization of a dye is rather time-consuming and the application of an unknown dye risky, the well-tried dyes and procedures are preferred. Thus, only a few compounds are systematically used for intracellular pH measurement, viz., fluorescein and its derivatives, pyranine, and derivatives of hydroxy-coumarin. Instead of choosing completely new dyes, chemical modifications of the existing, established dyes are tested, and they often turn out to be superior to the parent compound. Some newly synthesized compounds are also offered commercially by Molecular Probes[4] as a challenge for enterprising investigators. As the incubation of cells in the acylated form of the dye has become an established way of feeding cells with a pH indicator (trapped indicator), many of these dyes are already also offered in the acylated version.

B. Fluorescein and Its Derivatives

Fluorescein is a famous fluorescent dye with a long history of application in biological, chemical, and medical science. Its absorption and fluorescence properties are significantly pH dependent. There are nine different forms of fluorescein, six of them existing in aqueous solution,[5-7] with corresponding p*K* values 2.2, 4.4, and 6.7. As all of these six forms are absorbing and fluorescent (Figure 29), the resulting pH dependence of excitation and emission fluorescence spectra is rather complex.

In the physiological pH range of 6 to 7, the monoanionic and dianionic forms predominate; at lower pH the contribution of neutral molecules is apparent. The existence of two consecutive transitions (neutral molecule $\rightleftharpoons$ monoanion $\rightleftharpoons$ dianion) explains the exceptionally large range of pH response with consecutive contributions of excitation maxima at 435, 473, and 490 nm. The higher quantum yield of the cationic form relative to that of the neutral form allows for the extension of the applicability of 490/435 nm excitation ratio for measurement of pH values below pH 2. Therefore, fluorescein may serve as a pH indicator over the incredibly large pH range from 1 to 8, with all three transitions (cation $\rightleftharpoons$ neutral molecule $\rightleftharpoons$ monoanion $\rightleftharpoons$ dianion) involved.

The emission of all the six forms exhibits a maximum around 520 nm, making this wavelength ideal for monitoring of fluorescence intensity. The emission of cationic and dianionic forms decays more rapidly on the red side than that of neutral and monoanionic forms (Figure 30).

The extraordinary stability of p*K*, dependent neither on the composition of the medium[8-10] (cf. Figure 11) nor upon the excitation ($pK = 4.4$ does not change at all, $pK = 6.7$ shifts to 6.9),[5,6] makes this dye ideal for pH measurement, in contrast to other dyes which exhibit a $pK - pK^*$ shift of typically several pH units.

The six forms of fluorescein differ not only in their spectroscopic properties, but also in their membrane permeability, behavior inside the cell, etc. One should bear in mind that most of the literature data apply to the properties of monoanionic and dianionic forms as they are abundant at neutral pH.

The acylated form of fluorescein is colorless and nonfluorescent, and the incubation of cells in its solution is used as a sensitive assay for cell viability. Viable cells develop intense fluorescence as they are gradually filled up with intracellularly formed fluorescein. This phenomenon, termed "fluorochromasia" by Rotman and Papermaster,[11] and the above-

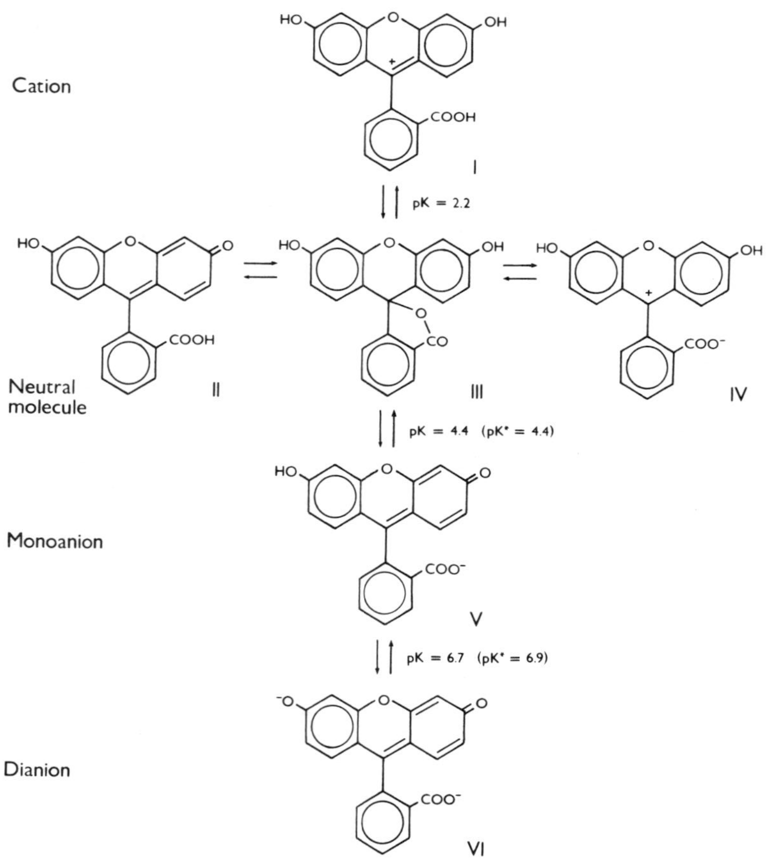

	ABSORPTION MAXIMUM [nm]	ε [M^{-1} cm^{-1}]	FLUORESCENCE MAXIMUM [nm]	γ [–]
Cation	435	55 000	520	0.9–1
Neutral molecule	435	16 000	520	0.2–0.25
Monoanion	435 and 473	30 000 and 31 000	520	0.25–0.35
Dianion	490	88 000	520	0.93

FIGURE 29. The dissociation scheme of fluorescein in aqueous solution and the basic optical characterization of its six forms. (Redrawn from References 5 and 6.)

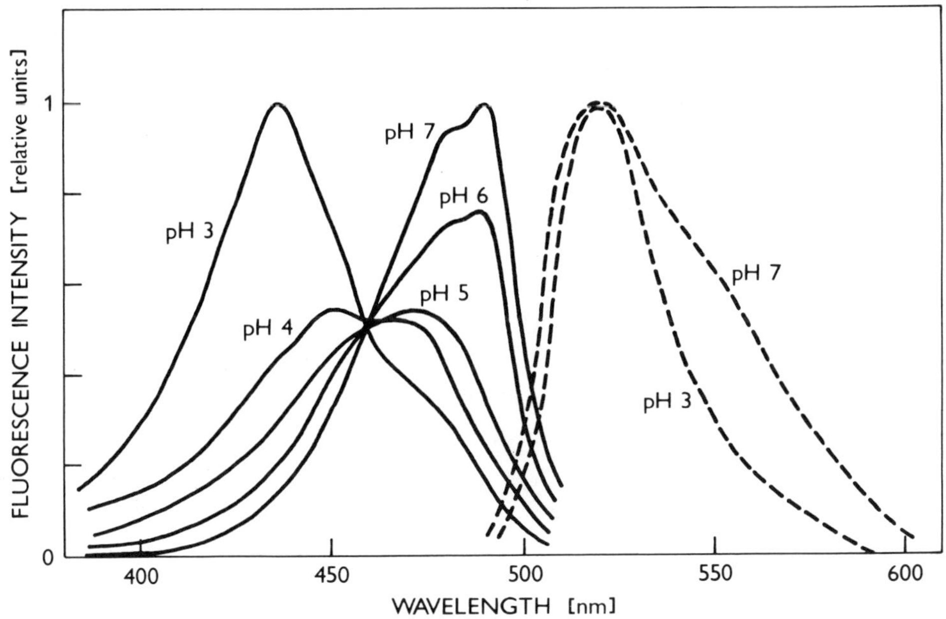

FIGURE 30. pH dependence of fluorescence excitation (full lines) and emission (broken lines) spectra of fluorescein.

mentioned pH dependence of fluorescence excitation, led to the idea of the employment of intracellularly trapped fluorescein for the measurement of intracellular pH by fluorescence.[12,13]

Various derivatives of fluorescein have been used such as 5- (and 6)-carboxyfluorescein (CF),[13-18] 2′,7′-*bis*-(carboxyethyl)-5-(and 6))-carboxyfluorescein (BCECF),[19-22,62] and 5- (and 6)-carboxy-4′,5′-dimethylfluorescein.[23-24] Due to lack of experimental data on the properties of the different derivatives and relative merits of each, it is difficult to judge which fluorescein compound is the best choice.[18] They vary in the ability to penetrate the membrane, p*K* stability, and stability of the acylated form. For instance, carboxyfluorescein has a lower leakage rate than fluorescein and supposedly does not penetrate inside mitochondria (in contrast to fluorescein[13]), probably because it has a higher electrical charge than the latter compound. The stability of acylated dimethylcarboxyfluorescein should be higher than that of carboxyfluorescein as both compounds, together with fluorescein diacetate, decompose spontaneously at pH above 6.5.[24,25]

All fluorescein derivatives (Figure 31) should principally follow the dissociation scheme of fluorescein (Figure 26), though with slightly different p*K* values. Similarly, their absorption, excitation, and emission spectra should resemble those of fluorescein, with slightly shifted maxima. Also, their acetylated forms are weakly absorbing and weakly fluorescent, similar to the colorless and nonfluorescent fluorescein diacetate.

Unfortunately, none of the fluorescein derivatives has a p*K* of the monoanionic-dianionic transition higher than 7, so that the use of fluoresceins for intracellular pH measurement remains limited to the acidic and neutral range.

Recently, azidofluorescein diacetate was used for photo labeling of intracellular proteins and subsequent measurement of intracellular pH.[26] In addition there is a fluorescein quenching rabbit antibody.[4]

C. Pyraninc

Pyranine (8-hydroxy-1,3,6-pyrene-trisulfonate), a derivative of pyrene, is an extremely pH-sensitive dye, convenient for the monitoring of intracellular pH (Figure 32).[27-36]

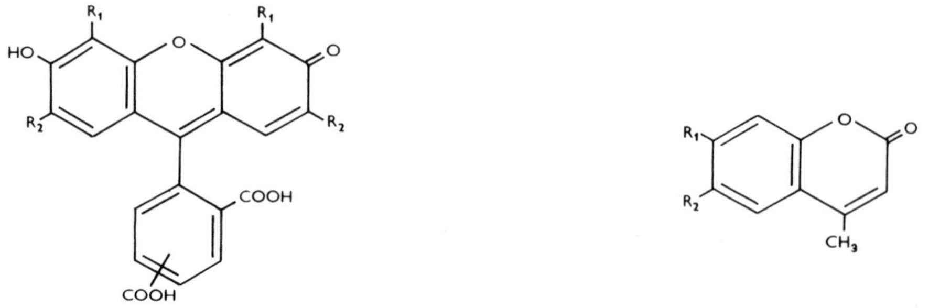

fluorescein $R_1 = R_2 = H$

5(and 6)-carboxyfluorescein $R_1 = R_2 = H$

5(and 6)-carboxy-4′,5′-dimethylfluorescein
$R_1 = CH_3$ $R_2 = H$

2′,7′-biscarboxyethyl-5(and-6)-carboxyfluorescein
$R_1 = H$ $R_2 = (CH_2)_2COOH$

7-hydroxy-4-methylcoumarin
$R_1 = OH$ $R_2 = H$

6,7-dihydroxy-4-methylcoumarin
$R_1 = R_2 = OH$

6,7-diacetoxy-4-methylcoumarin
$R_1 = R_2 = CH_3COO$

FIGURE 31. Derivatives of fluorescein and umbelliferone used in intracellular pH measurement.

Both forms of pyranine have the same shape of fluorescence emission spectra with a maximum at 510 nm. The pH-dependent excitation spectra show a 5000-fold change in the fluorescence intensity ratio after excitation at 400 nm (attributed to the trianionic form) and 450 nm (attributed to the tetraanionic form) in the pH range between 4 and 10. The ratio is apparently unaffected by the usual quenchers (1 *M* Cl^-, Br^-, or I^-, and oxygen).

Since the three sulfonate groups are completely ionized over the entire pH range from 4 to 10, the dye is effectively repelled from negatively charged surfaces and, consequently, attracted to the positively charged ones.[27] Furthermore, the high electrical charge of both forms of pyranine makes its p*K* value exceptionally vulnerable to ionic strength effects[27] in accordance with the theory. On the other hand, the membrane permeability of pyranine seems to be very low.

Pyranine is now commercially offered in the acylated form,[4] so that there should be no problem in bringing the dye into cells and having it trapped there. The fluorescence of pyranine can be quenched by *p*-xylene-*bis*-pyridinium bromide.[37,38]

D. Coumarins

A number of umbelliferone (7-hydroxycoumarin) derivatives have been used for intracellular pH measurements (Figure 31).

The 4-methylumbelliferone (7-hydroxy-4-methylcoumarin [4 MU]) was chosen for its relatively large change of fluorescence with pH,[39] suitable p*K* value ~7.8,[17,40] low toxicity, and uniform intracellular distribution. Both fluorescence excitation and emission spectra are pH sensitive (Figure 33).[18,40] Either the selective excitation (e.g., 365:334 ratio, monitored at around 450 nm)[17,31] or the ratio of emission intensities at 450 and 550 nm after 350 nm excitation can be used.[42,43] In both cases the ratio is an almost linear function of pH.

Similarly, 6,7-dihydroxy-4-methylcoumarin was used as an intracellular pH indicator.[44]

Both the above compounds are commercially available in acylated membrane-permeant versions.[45]

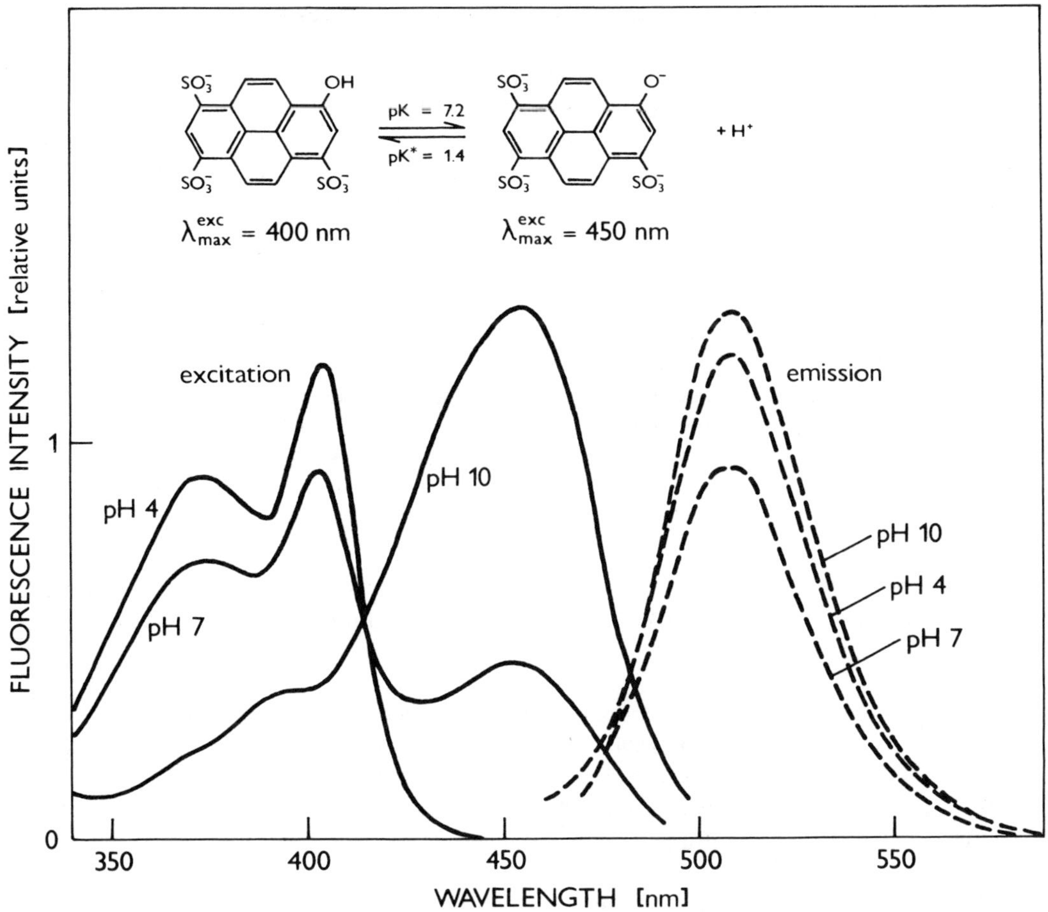

FIGURE 32. pH dependence of fluorescence excitation and emission spectra of pyranine. (Redrawn from Reference 27.)

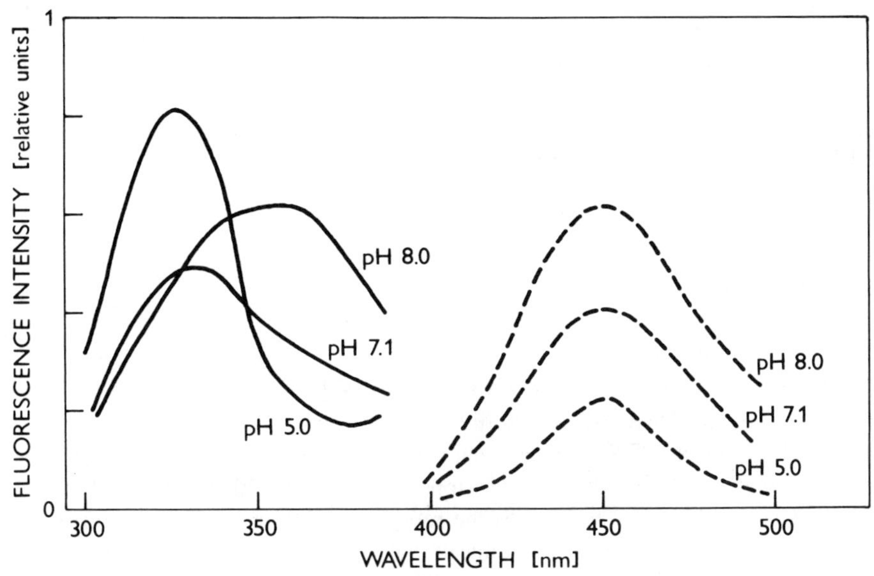

FIGURE 33. pH dependence of fluorescence excitation (full lines) and emission (broken lines) spectra of 7-hydroxy-4-methylcoumarin. (Redrawn from Reference 18.)

However, umbelliferone and, especially, its methyl derivatives are rather lipophilic and, hence, show a high rate of leakage,[18] significantly higher than that of fluorescein and its derivatives. This, together with high bleaching and UV excitation (which may be harmful for some more sensitive cells), is the weakest point of this dye.

The distinctly lipophilic versions of umbelliferone, viz., 4-heptadecyl-7-hydroxycoumarin[46] and 3-hexadecanoyl-7-hydroxycoumarin,[47] were used for monitoring of pH at the membrane surface.

E. Other pH-Sensitive Dyes

1. 1,4-Dihydroxyphthalonitrile

Its fluorescence emission spectra consist of two well-resolved peaks (one at 400 to 500 nm, the other at 500 to 600 nm), apparently corresponding to the dissociated and undissociated form of the dye, respectively. The ratio of these two fluorescences (excited between 340 and 380 nm) is pH-dependent.[48,49] The compound is available also in the acylated membrane-permeant version[4] and was used for intracellular-pH measurement of individual cells.[48]

2. Quinine

The extreme sensitivity of its fluorescence to Cl^- ions and ionic strength and a high pK − pK^* difference excludes this dye from biological applications[12,50] (pK_1 ~5.7, pK_2 ~9.9).

3. Quene 1

This dye is related to the tetracarboxylate Ca^{2+} indicator based on the quinolone chromophore, Quin 2. The cells are loaded after incubation in the tetraacetoxymethyl ester derivative which is hydrolyzed inside the cell to the intracellularly trapped tetracarboxylate anion. However, as chemically related to Quin 2, Quene 1 still retains a high sensitivity to Mg^{2+} and Ca^{2+} which makes it unsuitable for a reliable intracellular-pH measurement.[51-54]

F. Fluorescent Dyes Used in Distribution Techniques

Atebrin, methyl amine, acridine orange, and 9-aminoacridine are quite often used for the measurement of intracellular pH in distribution techniques as weak acids.[55] However, in contrast to other fluorescent dyes discussed earlier, both their excitation and emission spectra are pH-independent. The measurement of intracellular pH is based on the postulated (though theoretically unexplained and experimentally unverified) quenching of the fluorescence of intracellularly located dye. However, the observed fluorescence intensity hardly correlates with the $pH_{in} - pH_{out}$ values (with some exceptions in liposomes) and cannot be recommended for measurement in real cells.

III. EXPERIMENTAL ASPECTS

A. The Spectrofluorometer

The apparatus for fluorescence measurement consists of a light source used for the excitation of the sample, the sample, and a photodetector monitoring the light emitted from the sample (Figure 34). Excitation is usually provided by a xenon or mercury lamp or, in special cases, by a dye laser. Emission is detected mostly by photomultipliers, but also by special diodes, a photographic film, or a TV camera. In the usual geometrical arrangement, emitted fluorescence is detected at a right angle to the direction of the excitation beam in order to avoid the mixing of the excitation light with fluorescence.

1. Measurement in the Cuvette

The standard spectrofluorometer is usually equipped with two grating monochromators (one used for exciting light, the other for emission) and with a photomultiplier (Figure 35).

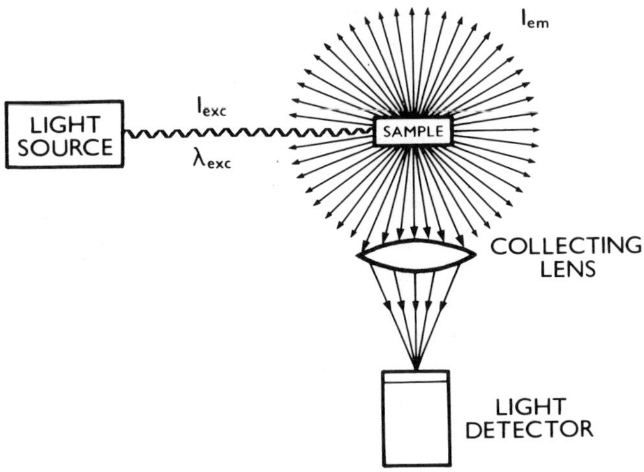

FIGURE 34. The principle of fluorescence measurement.

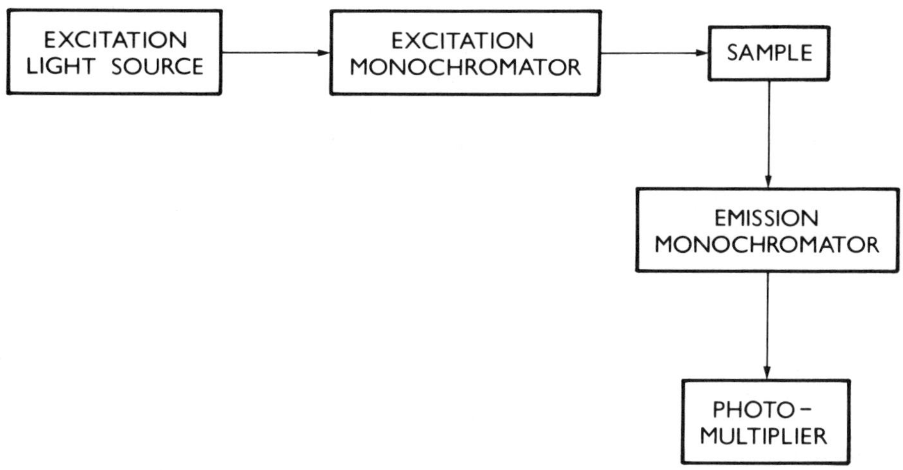

FIGURE 35. The basic schematic drawing of a spectrofluorometer.

It offers the possibility of measurement of steady-state fluorescence* excitation and emission spectra of a sample in a cuvette. The recorded spectra are displayed on a plotter, a printer, or a TV monitor, often after processing by a computer, which may correct the spectra with respect to the spectral sensitivity of the spectrofluorometer. Such spectra are called corrected excitation or emission spectra.

Spectrofluorometers are manufactured by many companies: e.g., Spex, Hitachi, Perkin-Elmer, Aminco SLM, Jobin Yvon, Shimadzu, Kontron, and Jasco. The basic apparatus costs $20,000 to $30,000 (U.S.). Differences between the various instruments are usually unsubstantial, though reliability, durability, and ease of handling may differ. The sensitivity (typically, 1 p*M* for strongly fluorescent and absorbing compounds, such as fluorescein or quinine) is virtually always sufficient, and it is often the background noise of the sample which limits the measurement. For this reason, some models have the excitation and emission

* Steady-state fluorescence distinguishes the "classical" fluorescence, obtained by continuous excitation, from time-resolved fluorescence.

monochromators doubled, which is convenient for the measurement of biological samples. However, a simpler and less expensive solution is represented by the insertion of a barrier or a narrow-band filter, which also decreases the contribution of false light. A typical volume of a sample is 2 ml (the inner dimension of a rectangular cuvette 10 × 10 mm), but there are special cuvettes for the measurement of volumes as small as 0.2 μl.

2. Microfluorometry

A fluorescence microscope offers an alternative to the classical spectrophotometer as the same type of measurements can be done with cells placed between microscopic slides in the fluorescence microscope instead of in a cuvette. The advantage of such an arrangement is the possibility to monitor pH values of a specific tissue section, of an individual cell, or of a subcellular compartment or area.

The principles of the microfluorometric arrangement are straightforward and simple (Figure 36). It is convenient to have the possibility of distinguishing the observed area by transmitted light inspection (bright-field or phase-contrast) in addition to the fluorescence observation. The light used here has a wavelength that does not pass to the film or photomultiplier (it is cut off by a barrier filter) and serves for visual observation of the sample only (Figure 37). The epifluorescence arrangement shown in Figure 36* is preferable, as it makes work easier (particularly with immersion oil) and gives both a higher spatial resolution close to the theoretical limit and a lower background fluorescence signal.

The selection of filters in fluorescence microscopy is critical. High-quality narrow-band interference filters should be employed for intracellular pH measurements (together with thermal filters, if necessary). Fluorescence is detected by a photographic camera (loaded with a highly sensitive film), a TV camera, or a photomultiplier which "sees" only a selected target area of the sample (spots down to 1 μm are obtained [Leitz]). An image intensifier may be useful.[56-58]

Adaptation of a commercially available microscope for the intracellular pH measurement is not difficult as many requirements of the pH techniques are already standard equipment or optional accessories of modern fluorescence microscopes (e.g., Leitz-Wild®, Nikon®, Zeiss®, Opton®, Jena®, and Olympus®). The way to transform a microscope into a microspectrofluorometer has been described in the literature many times.[9,59,60,73]

The installation of a unit maintaining interchangeability of excitation or emission filters is usually the most difficult point, together with the choice of a proper filter combination.

The sensitivity of microspectrofluorometry is usually at least two orders of magnitude lower than that achieved by ordinary cuvette spectrofluorometers, but it can be enhanced by dye lasers. However, as mentioned earlier, it is not the sensitivity but the background noise (background fluorescence of cells, medium, microscopic glasses, optics, filters, stray light, etc.) which sets the limit to biological experiments.

The application of fluorescence microscopy for the determination of *in vivo* pH maps of individual cells (intracellular pH topography) is described in detail in a separate Chapter (4D.V.).

3. Flow Cytometry

Individual cells passing in the form of a stream of drops (optimally, one cell per drop) through a certain "detection point" can also be investigated by fluorescence techniques, and their intracellular pH can be assessed. This is done in flow cytometry where the pH-sensitive dye, trapped inside the cell, is excited, and fluorescence is detected as the cell passes through a certain area. The usual arrangement allows one to measure the fluorescence

* Here, excitation light comes from above through the microscopic objective lens, and fluorescence is then collected by the same lens. No condensor is needed.

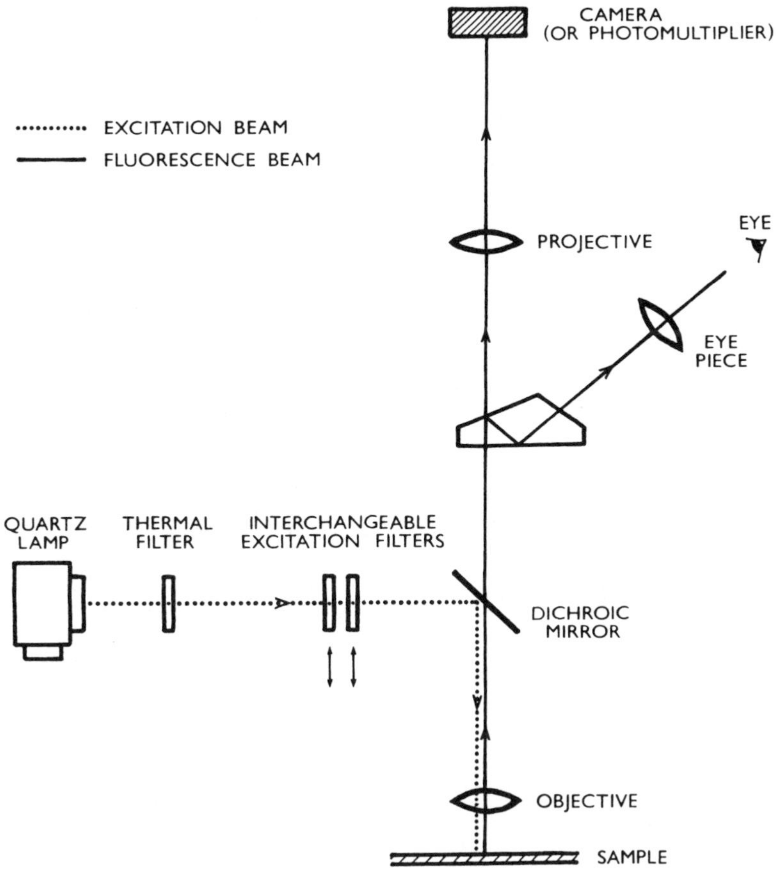

FIGURE 36. The principle of microspectrofluorometry (epifluorescence).

intensity at two different wavelengths (Figure 38). However, the use of two different light sources (e.g., two dye lasers) aligned close to each other allows the fluorescence signal at two different excitation wavelengths to be measured. Two consecutive fluorescence pulses appear in rapid sequences (Figure 39). Therefore, all principles of intracellular pH measurement described earlier in this chapter can be applied to flow cytometry.

Flow cytometry yields information about individual average pH values of a large number of cells (typically, 5000 cells per second) and offers the possibility to differentiate between small subpopulations of cells. The technique is also discussed in more detail in a separate Chapter (4D.IV.).

B. Techniques of Measurement

In comparison to NMR and optical absorption, fluorescence techniques offer a larger number of parameters for measurement and, therefore, may look complicated. However, the opposite is true. If a senior person with good experimental knowledge is consulted, any individual can get down to work with fluorescence almost immediately after reading the spectrofluorometer instruction manual. The measurement of steady-state fluorescence excitation and emission spectra is, in principle, very simple and straightforward. The measurement of time-resolved fluorescence is, on the other hand, rather difficult, and its complexity is comparable to, if not exceeding, that of NMR spectroscopy.

The major problem with biological experiments is the signal-to-noise ratio, viz., the contribution of background fluorescence and scattered and stray light (Figure 40). Their intensity can be diminished in the ways described below (cf. Figures 41 and 42).

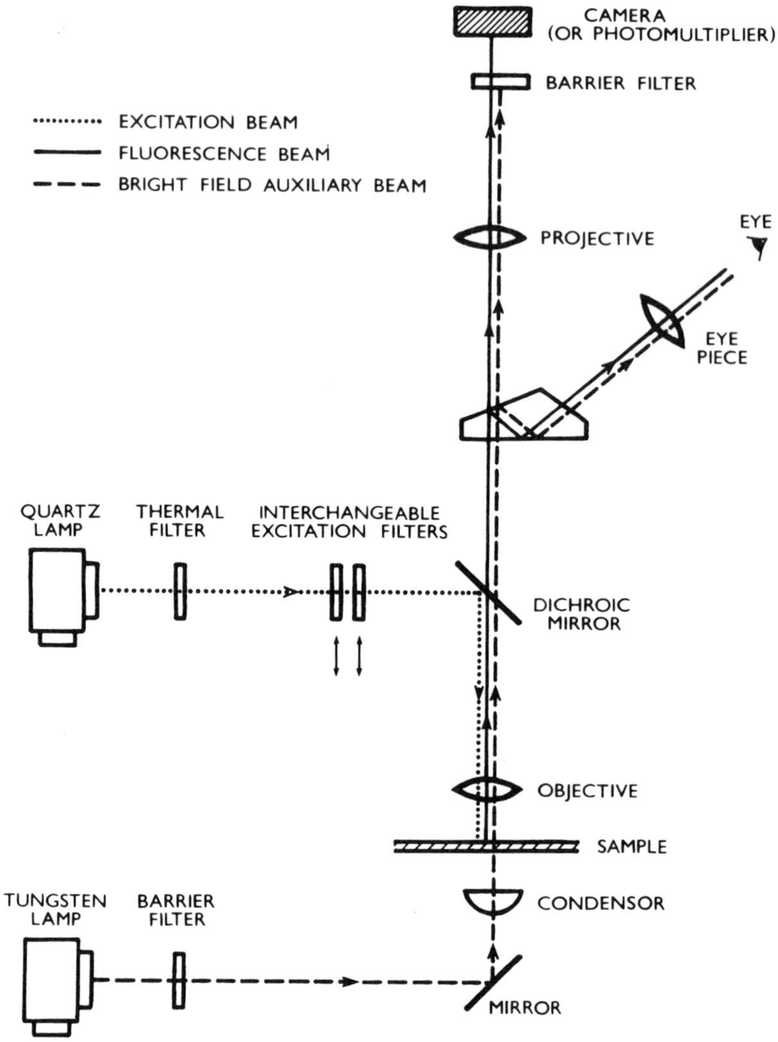

FIGURE 37. Combination of microspectrofluorometry with a simultaneous bright-field visual observation.

Background fluorescence can be suppressed if the optimal (compromise) combination of excitation and emission wavelengths is used, as both the background signal and the fluorescence of the pH-indicating dye depends on the excitation wavelength. The importance of a detailed knowledge of excitation/emission properties of the dye, of the cells alone, and of cells with the dye (internal dye has similar, but not necessarily equivalent, spectral properties) cannot be overemphasized. Sequential scanning of emission spectra with the excitation wavelength gradually increasing (let us say, in 5-nm steps) may, indeed, be very useful (Figure 43).

Scattered excitation light may appear quite strong in measurements of turbid biological samples and must be eliminated (e.g., by insertion of a cut-off barrier filter) which does not allow the excitation light to enter the photodetector (Figure 42).

Raman scattering appears as a strong peak, red-shifted from the excitation wavelength. The position of the peak shifts in accordance with the shift of the excitation wavelength (Figure 44) (on the frequency scale, their difference is fixed), so that the possible coincident overlapping of the Raman band with the fluorescence of the dye may be eliminated. Raman

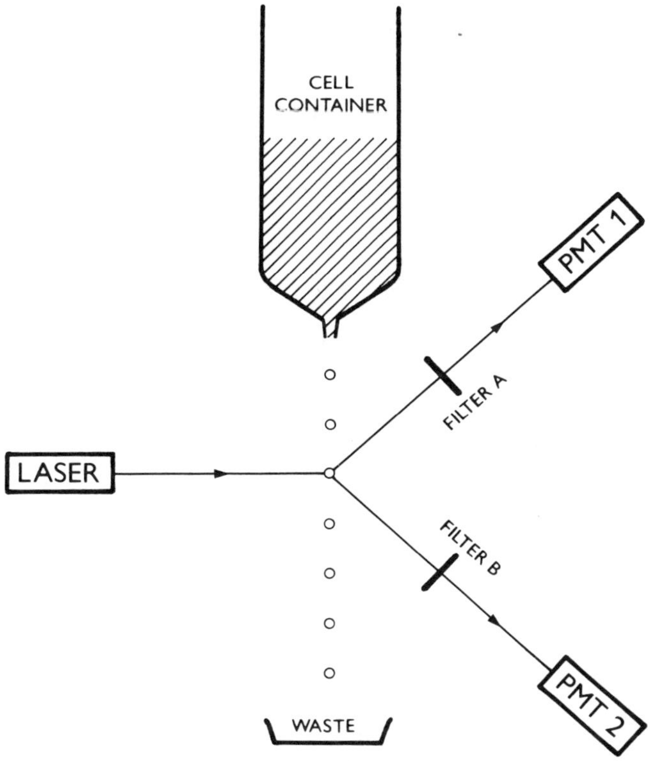

FIGURE 38. The principle of flow cytometry applied to the measurement of average intracellular pH in individual cells using the double-emission-wavelength technique.

scattering arises from the solvent in the cuvette (water, in biological samples, and is polarized (the excitation light leaking out of the grating monochromator is usually considerably polarized), so the insertion of a polarizer may also help.

Stray light is the light coming from the monochromator at wavelengths different from the desired excitation wavelength set on the spectrofluorometer. It consists of the light of virtually all wavelengths emitted by the excitation light source. In grating monochromators, stray light is dominated by the second harmonic, i.e., light with wavelength equal to half the wavelength set. Stray light is usually weak and may be neglected, but in strongly scattering samples its intensity may reach an unacceptable level. In such cases the use of a spectrofluorometer with double monochromators or the already-mentioned insertion of a narrow-band interference filter can help considerably and make the fluorescence measurement feasible.

It may be surprising how stray and false light finds its way into the photodetector. This may happen if the light strikes upon the interference filter at a low angle because the transmission properties of the interference filter depend on the angle of incident light.

In fluorescence microscopy and flow cytometry, it is desirable first to test the dye in the medium and the dye inside the cell using a reliable spectrofluorometer equipped with monochromators (possibly even time-resolved spectra). These introductory control experiments should be carefully performed as any error can later be paid for dearly.

C. Sensitivity, Temporal, and Spatial Resolution

The sensitivity of a spectrofluorometer is so high (~1 p*M*) that it can hardly be the limiting factor in biological experiments. Photomultipliers operate close to the theoretical limits.

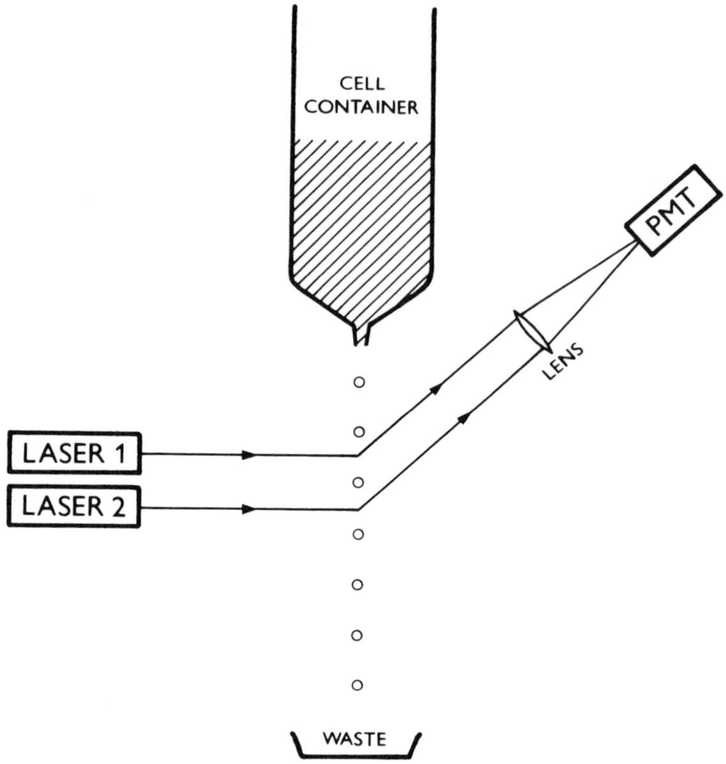

FIGURE 39. The principle of flow cytometry applied to the measurement of average intracellular pH in individual cells using the selective double excitation of the dye.

Often they really respond to individual photons striking upon the photocathode (the so-called "photon-counting" apparatus). If necessary, the intensity of fluorescence can be increased by special accessories (mirrors, collecting lens with a large aperture) or by using a more intense excitation source (e.g., a dye laser in microspectrofluorometry). However, the use of higher intensity of excitation light can make the bleaching of the dye significant and in this way complicate the fluorescence measurement (significant in fluorescence microscopy).

The spatial resolution is limited by theoretical limits imposed by the wave nature of light. In an epifluorescence microscope the definition actually achieved is close to the theoretical limit of 150 to 200 nm.

The temporal resolution of intracellular pH changes is, in practice, limited by the relaxation time needed for a pH-indicating dye to reestablish an equilibrium with the protons of the medium (reportedly ~30 μs) [61] rather than by the resolution of the fluorescence technique (limited by the fluorescence lifetimes of some 10 ns). However, in routine instruments, the temporal resolution is limited by the time constant of the detecting electronics. With improved electronic detection, the response of the photomultiplier (~1 ns) may become limiting. Therefore, fluorescence techniques are ideal for monitoring fast pH changes. Similar to the absorption technique, their temporal definition is limited by the response of the dye.

D. How to Prepare a Biological Sample

1. Introduction

Visible and near UV light is virtually harmless to cells at the intensities used in spectrofluorometry. Furthermore, fluorescent dyes, if not clearly toxic, cause little if any damage

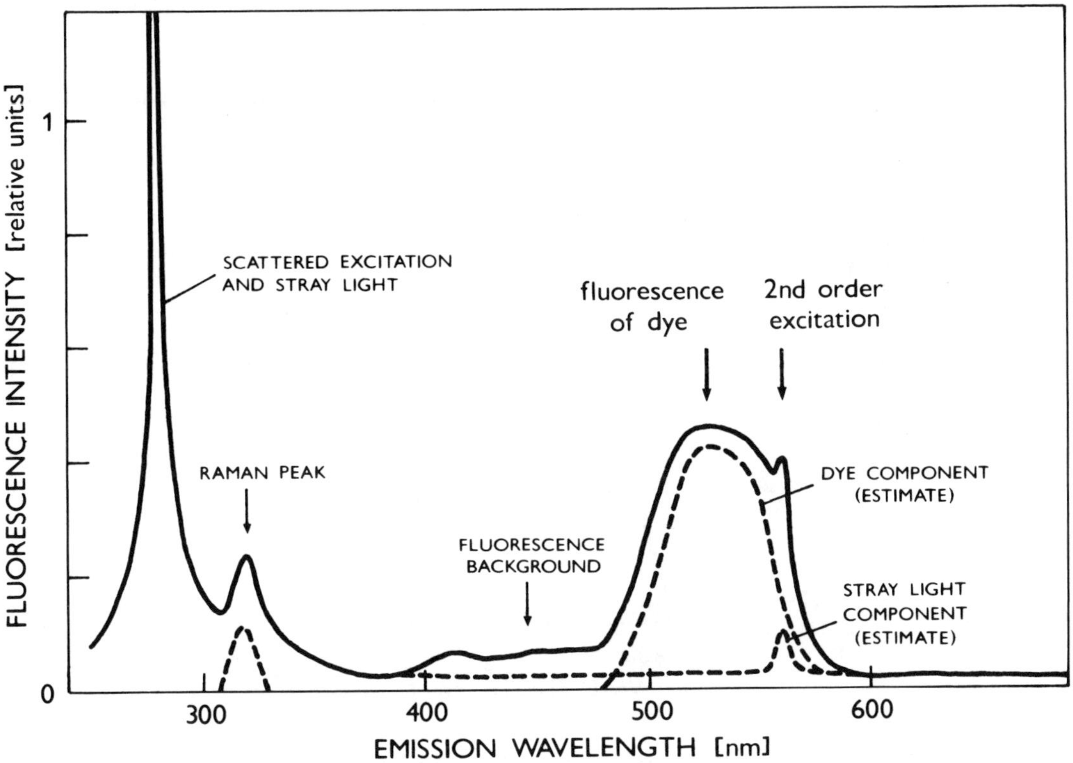

FIGURE 40. An example of an untreated fluorescence emission spectrum.

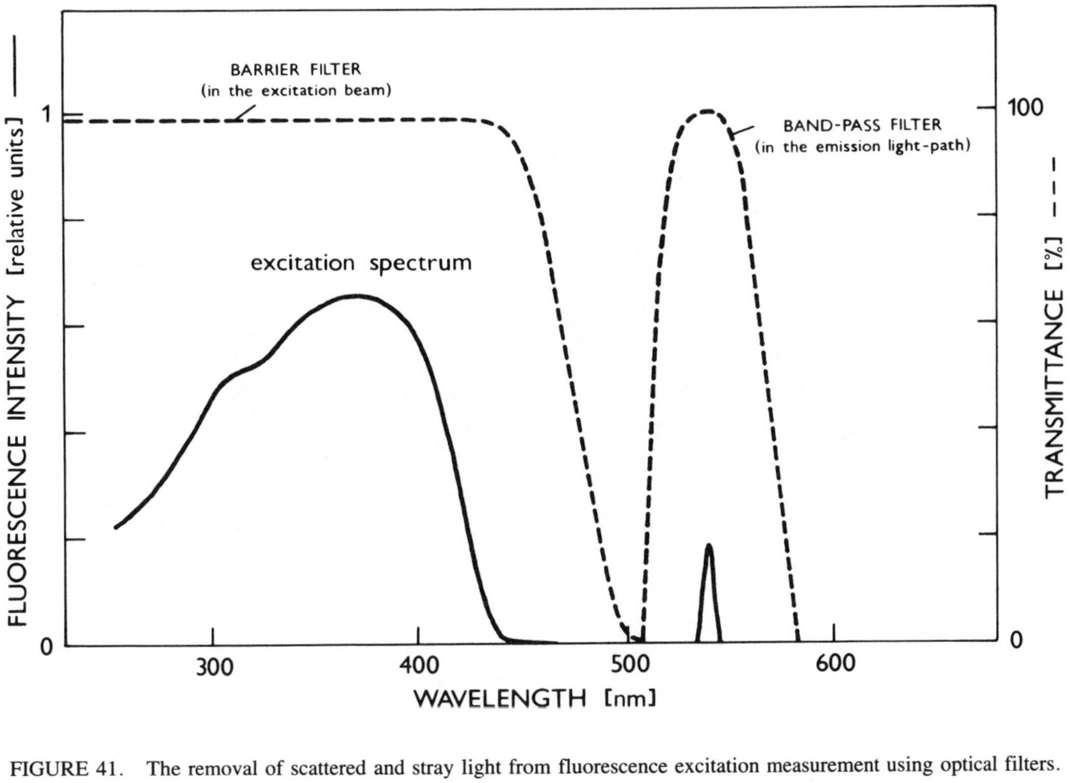

FIGURE 41. The removal of scattered and stray light from fluorescence excitation measurement using optical filters.

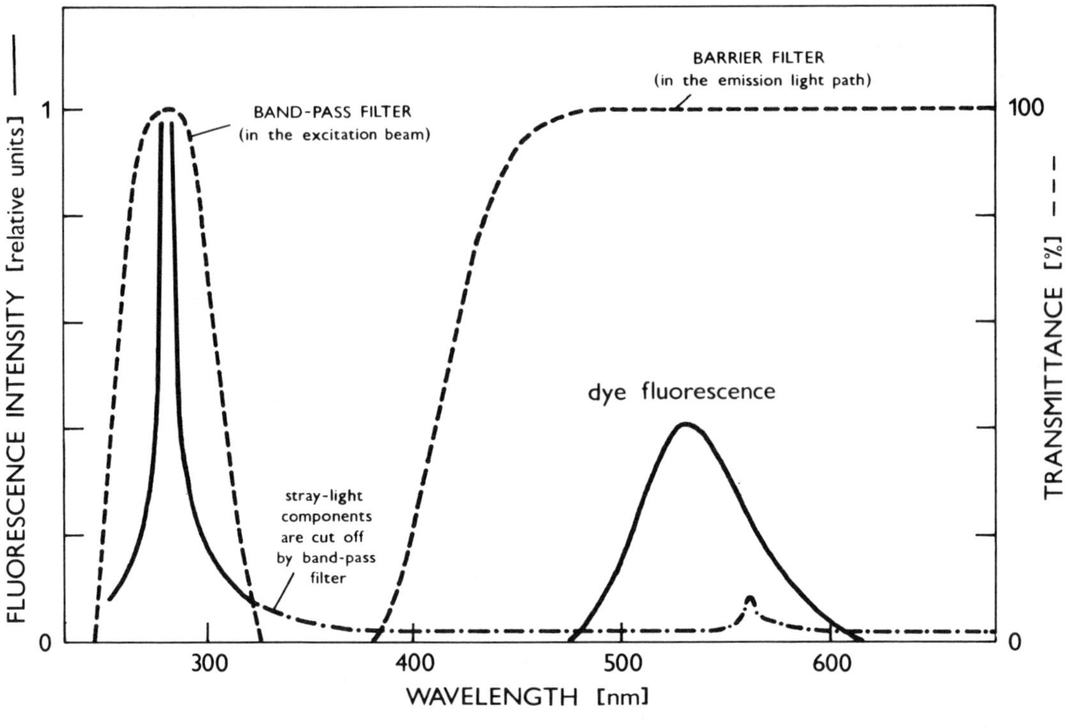

FIGURE 42. The removal of scattered and stray light from the fluorescence emission measurement using optical filters.

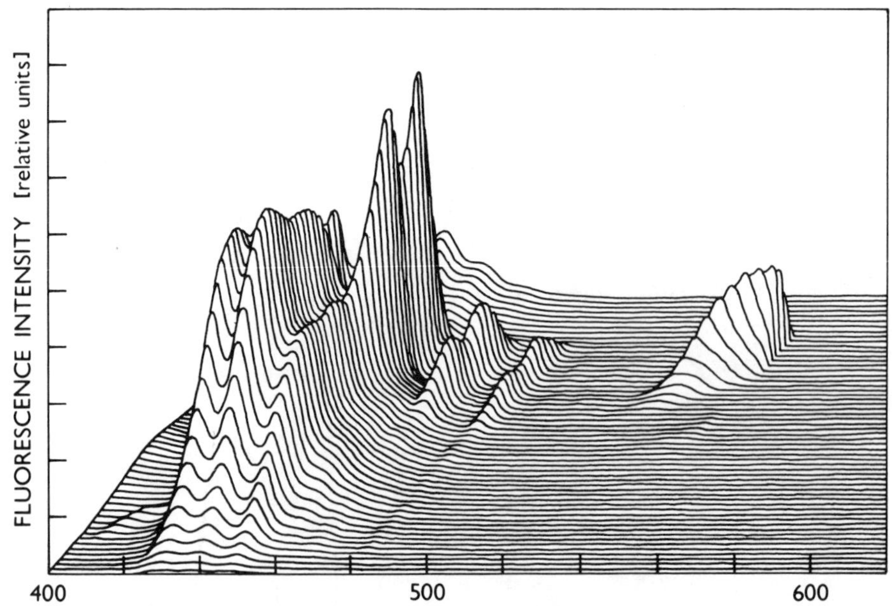

FIGURE 43. Sequential scanning of emission spectra at a gradually increasing value of excitation wavelength (a pseudo-three-dimensional plot).

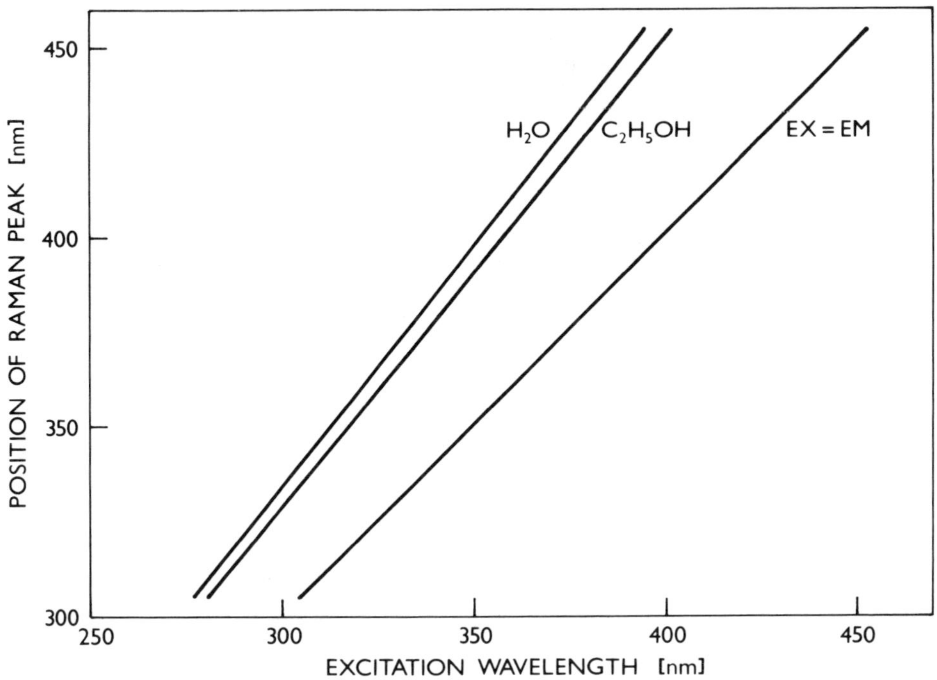

FIGURE 44. Dependence of the position of the maximum of the Raman peak in water and ethanol on excitation wavelength. The EX = EM line represents a reference with no shift.

to the cell because of the really low concentrations used. For instance, the viability of cells can be confirmed by double staining when vital stains are used to demonstrate the harmlessness of the pH-sensitive dye.[42] Furthermore, the filling of cells with fluorescein, following incubation in fluorescein diacetate, was actually introduced as a vital staining technique distinguishing between living and dead cells.[11]

The following section deals mainly with the measurement of cells in a cuvette. Microscopic measurements use standard approaches of optical microscopy. The flow-cytometric procedures can handle any suspension of cells in the recommended concentration range of 10^5 to 10^7 cells per milliliter and with cell sizes of 0.5 to 100 μm.

The preparation of a sample for fluorescence measurement should be preceded by washing the cells to remove any fluorescent metabolites in the growth medium (the presence of which could contribute to the fluorescence background). Introduction of the dye into the cell can be achieved as described in Chapter 4A.

2. The Cell Concentration and the Geometry of Measurement

The concentration of cells is closely related to the choice of the geometrical configuration of the measurement.

Dilute suspensions are measured in the same geometrical arrangement as solutions, in a right-angled geometry. The high sensitivity of the fluorescence technique allows measurement of very dilute samples which appear clear when viewed with a naked eye.

Turbid suspensions must be measured in the frontal light arrangement (Figure 45). For such a measurement, a special adapter is needed, such as the solid/powder sample holder accessory in which the cuvette is fixed at the place of a solid sample.

The frontal illumination (also called front-face viewing) is convenient not only for turbid samples, but also for the measurement with cells requiring attachment to a substratum (Figure 46).[1,62] Nevertheless, it should be emphasized that the use of frontal light geomtery signif-

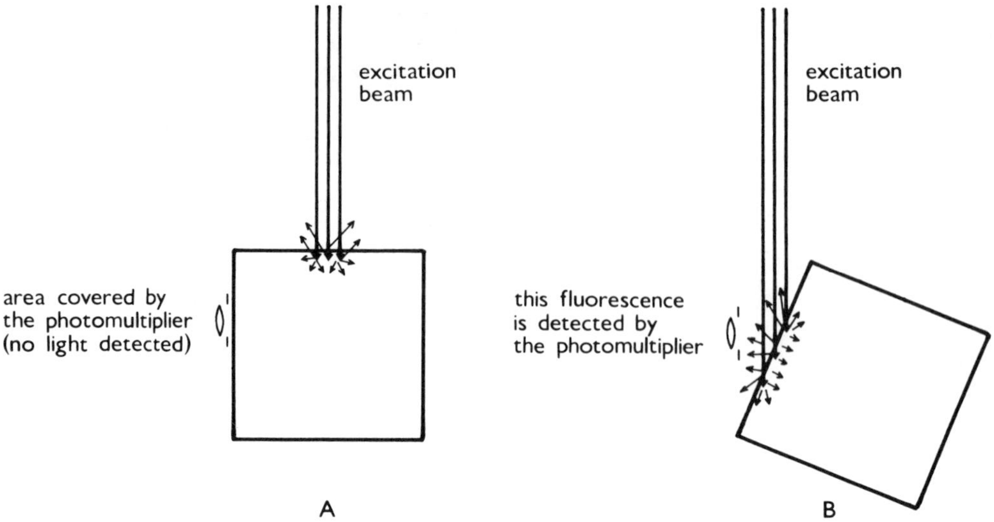

FIGURE 45. The frontal light arrangement of fluorescence measurement. With highly turbid and/or highly absorbing samples, only a thin layer close to the cuvette wall is excited. In the usual right-angled arrangement (A), the fluorescence cannot reach the photodetector, but it can do so in the frontal one (B).

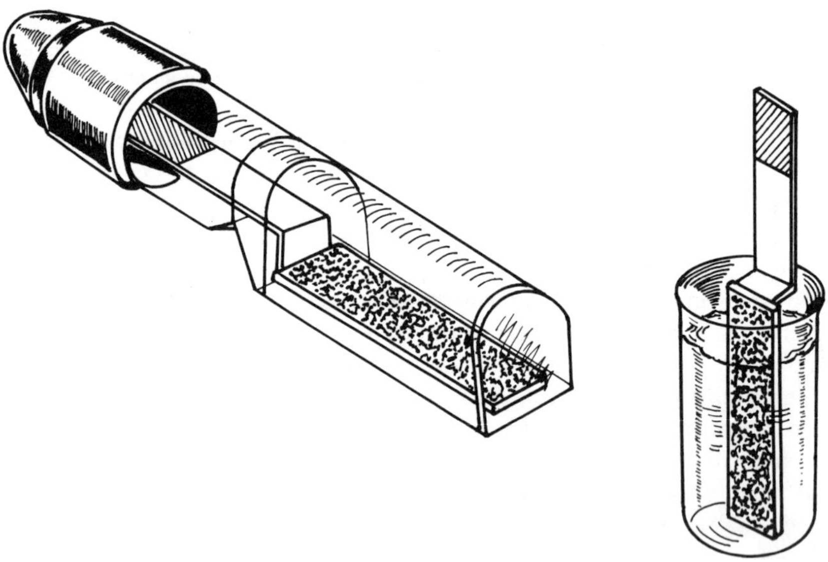

FIGURE 46. A schematic view of cells growing on plastic slides which can be inserted into the spectrofluorometer cuvette and used for intracellular pH measurement. (Redrawn from Reference 13.)

icantly increases the danger of contribution of false light to the fluorescence. Therefore, the use of narrow-band interference filters and cut-off band pass or high-pass filters inserted into the excitation or emission beam is inevitable.

3. Stirring

Sedimentation of cells causes the intensity of the fluorescence signal to diminish gradually. Therefore, suspended cells larger than 3 to 5 μm should occasionally or continually be

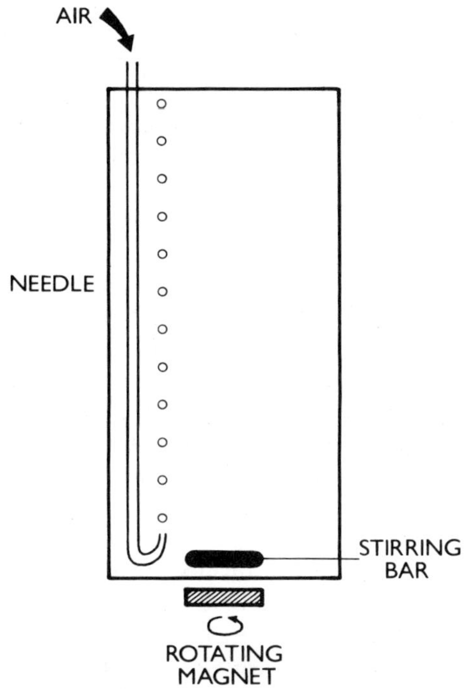

FIGURE 47. Stirring and aeration of a cell suspension in a cuvette.

stirred (Figure 47) by any technique that does not interfere with the passage of excitation and emission beams. Stirring can also be substituted by gently bubbling with air or, better, with an inert gas.

4. Aeration and Nutrition Supply

The requirement of oxygen and nutrient supply in fluorescence measurement is much less critical than in NMR because the densities of cell suspensions used are significantly lower. Almost any kind of aeration and injection of nutrients can be effected in the cuvette, provided it does not cross the path of the excitation and emission beams (Figure 47). Any increase of noise can be eliminated by an electronic averaging of the signal (larger time constant).

5. Temperature Control

A thermostated cuvette compartment is an optional accessory of almost every commercial spectrofluorometer, although it usually allows only for measurement in a right-angled geometry.

E. Interpretation of Results

The first step in the interpretation procedure is a thorough comparison of the excitation and emission spectra (and possibly the absorption spectra and fluorescence lifetime) of the sample with that of the dye in buffers. This confirms not only that one, indeed, observes the fluorescence of the dye one would like to observe, but it is also essential for the estimation of the behavior of the dye inside the cell (e.g., extent of binding; cf. Chapter 4A).

Fluorescence background arising from the spectrofluorometer itself (false fluorescence emitted by optical components, stained cuvettes, fingerprints, cellulose dust, etc.) can be subtracted from the signal. However, the natural fluorescence of cells without a dye (arising, e.g., from porphyrins, flavins, nucleic acids, proteins, etc.) cannot simply be subtracted from the total signal as it is often quenched upon addition of the dye. (There is an excitation energy transfer between the above-mentioned intrinsic chromophores and the dye molecules.) Therefore, it is recommended to increase the internal concentration of the dye and optimize the excitation wavelength to make the background signal negligible.

F. Problems of pH Calibration

1. Specific Aspects of Fluorescence Techniques

A careful, methodical approach to fluorescence calibration is always rewarding. Each dye or, better, each combination of a dye with cells should be thoroughly tested. The calibration procedure should generally follow the rules described in Chapter 4A. However, there are a few aspects specific for fluorescence techniques summarized below such as impurity of the dye, photochemical instability of the dye (e.g., photobleaching), a high fluorescence background of cells, and quenching of the fluorescence of the dye.

The dye should be purified chromatographically, and then the right concentration, the correct intensity of excitation, and an optimal excitation-emission wavelength combination, must be chosen.

In contrast to NMR spectroscopy, the formation of dimers, excimers, aggregates, etc. usually presents no serious problem to pH calibration and often can be neglected. The only problem, that may distort the pH measurement in a way comparable to the effect of divalent cations and basic proteins in NMR, is the sensitivity of some dyes to fluorescence quenching. For instance, the quenching of quinine fluorescence by Cl^- excludes its use as a pH probe in biological systems. The quenching is specific for each dye and, thus, it is useful to check the quenching effect of all substances, the effect of which on the intracellular pH is studied (namely, cations and colored substances) prior to pH measurement, even if the concentration added is minimal. On the other hand, quenching of leaked extracellular dye may be quite useful (cf. section on leak in Chapter 4A).

In the absence of quenching, the reliability of the pH response of the dye can be distorted only by shifts in the apparent pK as described in Chapter 4A. According to the Debye-Hückel theory, both the pH and the ionic strength of the sample should influence the ratio between the two forms of the dye. However, some dyes seem to be almost insensitive to the ionic strength of the medium as monitored by the fluorescence signal in the way described; some are more sensitive.

The difference is pK' values between the ground state and the excited state should not influence the pH response as the response of the dye to the ambient pH should occur on a microsecond scale, while the fluorescence lifetimes are on a nanosecond scale. Therefore, no redistribution between the basic and the acidic form of the dye, caused by an altered pK' value during the existence of the excited state, should take place. However, in some dyes, such processes (apparently controlled by diffusion because their effect diminishes with increasing viscosity of the medium) were experimentally observed in this laboratory.[63]

2. Specificity of the pH Calibration Curve

The intensity of the excitation light from the light source and the sensitivity of the light-detecting part vary with the wavelength in each fluorescence apparatus in a specific way. Thus, the shape of excitation and emission spectra recorded by the spectrofluorometer is distorted, and the standard pH curves derived from such spectra are specific for each type of spectrofluorometer. Similarly, they are specific for each filter combination in fluorescence microscopy.

However, if standard curves of a particular substance are plotted on a log scale, they all have the same shape, the differences between the different instruments appearing as a shift in the direction of the y-axis (Figure 48). Modern spectrofluorometers are often equipped

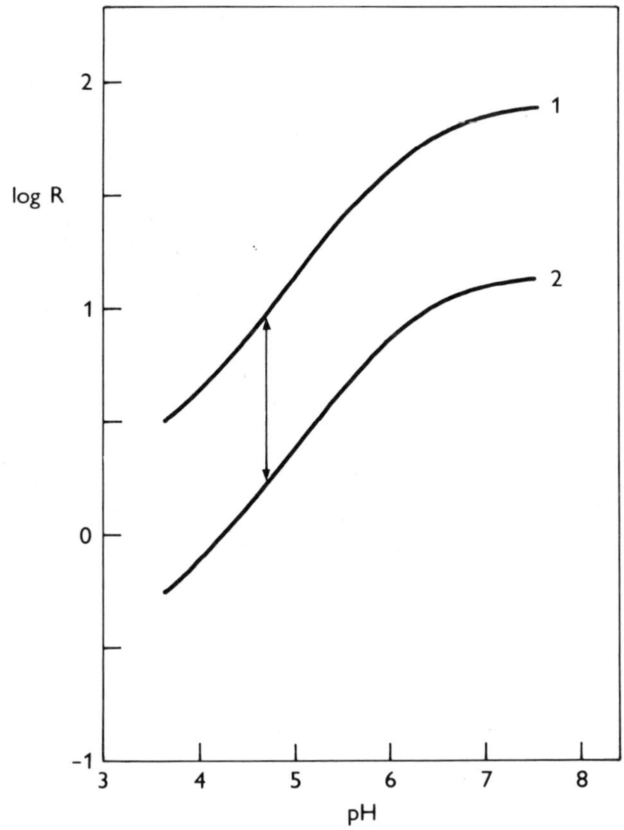

FIGURE 48. Comparison of pH standard curves obtained on the basis of spectrally uncorrected fluorescence spectra from two different spectrofluorometers (1 and 2).

with a computer program which corrects for these effects and yields the so-called corrected fluorescence spectra. The calibration curves based on such corrected spectra are then independent of the apparatus used.

G. A Typical Experimental Result

The following excitation fluorescence spectrum was recorded (Figure 49A) in determining the intracellular pH of the yeast *Saccharomyces cerevisiae*.

The conditions were as follows:

1. Freshly harvested cells were washed twice and resuspended in the cuvette to a final density of 5 mg dry weight per milliliter.
2. No stirring or bubbling was used.
3. The emission wavelength was set at 520 nm.
4. Cells were filled with fluorescein formed *in situ* using a 3-min incubation with 0.1 m*M* fluorescein diacetate.
5. A frontal geometry adapter for measurement of solid samples (accessory equipment with ordinary spectrofluorometers) was used.
6. A narrow-band interference filter with a peak transmission of 520 nm was inserted in front of the emission monochromator slit.
7. The background signal was neglected.
8. Calibration of the pH curve was derived using buffers.

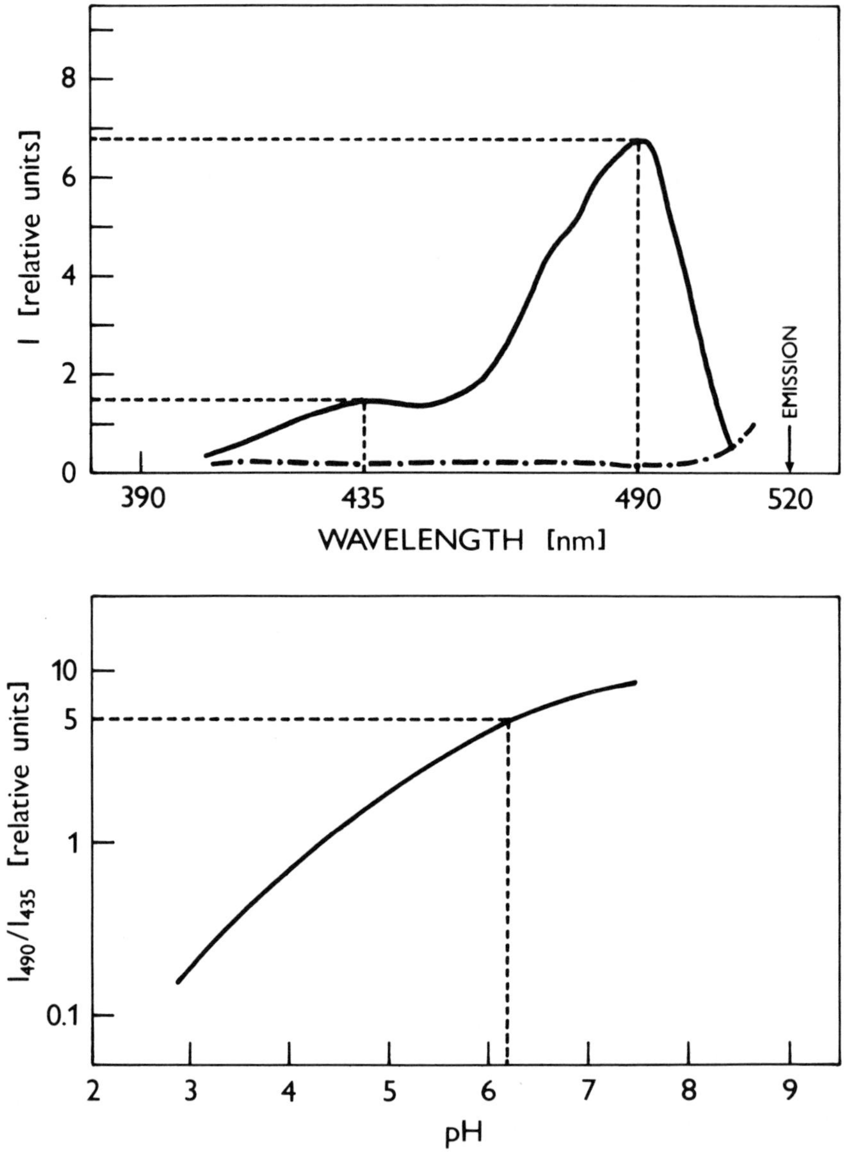

FIGURE 49. A typical result of fluorescence measurment of intracellular pH (above) and the corresponding pH standard curve (below). For details see the text.

Using a standard pH curve (Figure 49B), the intracellular pH was estimated to be 6.2.

IV. FLOW CYTOMETRY AND CELL SORTING

A. The Flow Cytometer

Flow cytometry offers the possibility to obtain easily quantitative data about cellular parameters of a large number of cells in a short time. The technique can register both light-scattering and fluorescence data, so that fluorescent pH indicators trapped inside cells provide individual intracellular pH values of many thousands of cells. Therefore, the pH values of various cell subpopulations within a studied cell suspension can be monitored. Both principles of fluorescence intracellular pH measurement, viz., selective excitation of two forms of the dye or their difference in emission spectra, can be employed (Figures 38 and 39).

A commercial flow cytometer (often functioning as a cell sorter) costs some $50,000 (U.S.) (e.g., EPICS® made by Coulter® Corporation or FACS® made by Becton Dickinson). The basic principles of flow cytometry and cell sorting have been quite recently reviewed in detail.[64-67]

A flow cytometer consists basically of three or four main units:

1. A fluidics system delivering the sample (flow chamber)
2. An optical system for examining the sample (light source and detectors)
3. A computer for data analysis
4. In addition, a flow cytometer possibly equipped with a device for sorting of cells

The purpose of the fluidics system is to create a stream of single cells which are linearly aligned and pass sequentially through the detection system.

The optical system uses a laser light source and several optical detectors. Each cell leaving the flow chamber and entering the liquid jet will interact with the laser beam and produce a fourfold signal:

1. A forward light-scatter signal (related to cell size)
2. A right-angle light-scatter signal (related to intracellular structure)
3. A fluorescence signal in wavelength band A
4. A fluorescence signal in wavelength band B

For intracellular measurement, the first two parameters can be used for "gating", i.e., a selection of a homogeneous population of cells within a heterogeneous cell mixture; and the third and fourth signal, for the characterization of intracellular pH.

Thus, with pH-indicating dyes exhibiting a pH-dependent emission spectrum, an intracellular pH measurement is easy to accomplish. The wavelength band A is chosen to correspond to the emission of the acidic form, and the wavelength band B to the emission of the basic form of the dye.

The situation is slightly complicated if dyes with pH-dependent excitation spectra (such as fluorescein) are employed. In such a case, a second dye laser must be used. The light beam of this second laser is aligned so that its intersection with the liquid stream is located some micrometers downstream compared to the first laser. The fluorescence light pulse produced by the cell interacting with the second laser beam is delayed by the time needed by the cell to travel the distance between the two laser beams and can be distinguished electronically from the former pulse. Assuming that the first dye laser excites the acidic form of the dye and the second laser excites the basic form, a comparison of the two signals gives the value related to intracellular pH.[42,43,48,68,69]

B. A Typical Experimental Result

Gerson[42] used 4-methyl-umbelliferone, which exhibits pH-dependent emission spectra and is, thus, easier to use in flow cytometry. He showed that stimulation of lymphocytes splits the originally homogeneous cell culture into two subpopulations with different pH values (Figure 50).

V. INTRACELLULAR pH TOPOGRAPHY

A. Introduction

A time-resolved, three-dimensional map of intracellular pH of a living cell represents the ultimate of intracellular pH measurement. Determination of the pH topography of a free-living single cell, based on a computerized image analysis of fluorescence microscopic

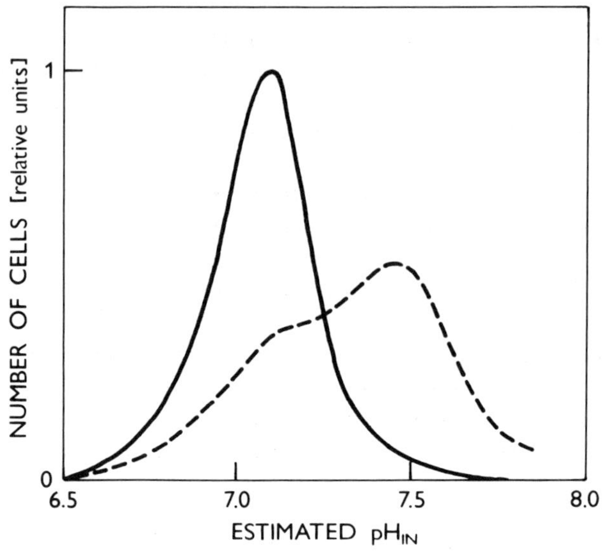

FIGURE 50. Distribution of intracellular pH values of individual cells before (full line) and after (broken line) a stimulation, as revealed by flow cytometry. (Redrawn from Reference 42.)

images, fulfills a large portion of this vision. The technique was developed in our laboratory in 1983[70] and represents a natural extension of a classical "cuvette" fluorescence pH-measuring technique developed a year earlier.[12] The combined use of fluorescence microscopy and intracellularly "trapped" fluorescein seemed to be ideally suited for estimations of local intracellular pH values with a spatial resolution of up to the definition limits of an optical microscope. The possibility of construction of such pH maps and their confrontation with cellular morphology was both a temptation and a challenge.[9]

The technique as described below gives two-dimensional pH maps of a cell (and of its immediate neighborhood, if desired; see pH profiles, Section D.) *in vivo*, with a spatial definition of 150 to 200 nm, a pH resolution better than 0.05 pH units, and a temporal resolution of 20 to 40 s (limited in our case by the speed of the film used).

B. The Principle

The intracellular pH topography technique starts from the technique of selectively excited acidic and basic forms of intracellularly trapped fluorescein (cf. example on Figure 49). It is based on the following principal steps (Figure 51):

1. Filling of cells with a pH-sensitive dye
2. Taking two microphotographs in sequence of a particular cell using selective excitation to excite the acidic and the basic forms of the dye
3. Computerized digital-image processing of both pictures and calculation of local differences in densities
4. Turning of these values into local pH values using a pH standard curve
5. Display of local pH values in the form of pH maps.

These steps are described in detail below.

1. Filling of Cells with Fluorescein

The trapping of *in situ*-formed fluorescein from its diacetylated precursor has become a

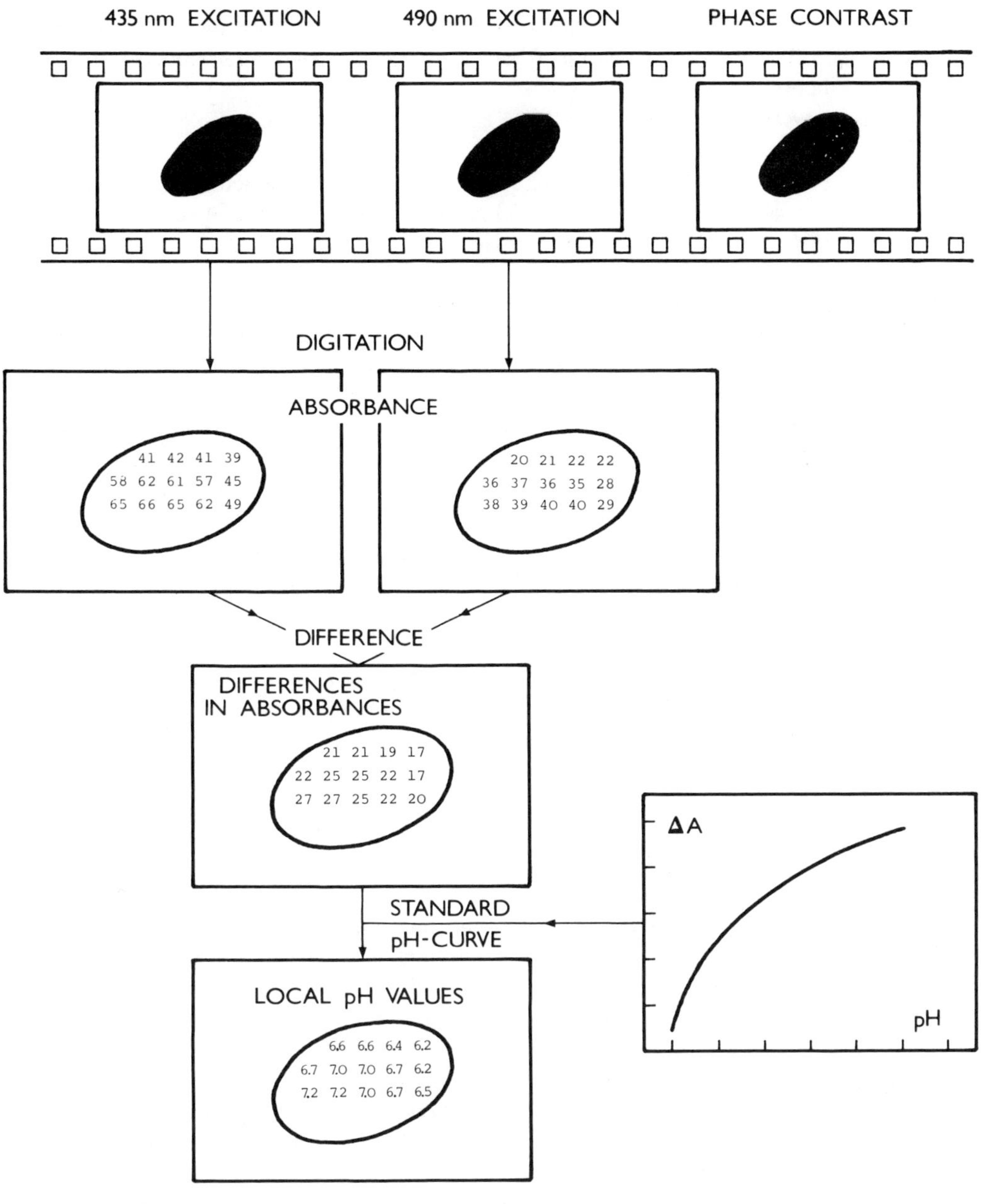

FIGURE 51. A schematic drawing of the procedure of intracellular pH topography determination.

very popular technique since the time of our pioneering experiments (discussed in Chapter 4A).

2. Microphotography

Routine microscopic techniques were used with the following technical details.

Epifluorescence was preferred for its higher spatial resolution and lower fluorescnece background.

One fluorescence picture was taken after excitation around 435 nm to excite the acidic form of the dye, the other with excitation around 490 nm to excite the basic form of the dye (both pictures exposed with an equal exposure time!) (cf. Chapter 4D.V.F.2). The third, phase-contrast bright-field, picture was taken to show the actual position and morphology of the cell. The fluorescence excitation filters were narrow-band interference filters used together with thermal filters.

High-sensitivity films with prolonged (forced) development were employed. Graininess of the film was of no concern as the resolution of the film was not limiting, and the fog of the film could easily be subtracted indigitized computerized densitometry. The speed of the film reached 39 to 42 DIN or 6,000 to 12,000 ASA (Ilford HP 5, developed in Microphen® for 30 min at 25°C), and the required time exposures were 10 to 60 s.

3. Computerized Densitometry and Digital Image Processing

The local densities of the individual spots of the photographs were determined by an image analyzer apparatus (TAS Leitz, W. Germany) and then rectified by the computer with respect to the real densitometric properties of the film (actual density curve, Schwarzschild effect, cf. Chapter 4D.V.F.2) and compared with the densities of the same spot (domain) in the second photograph. Since the optical density of the film is approximately directly proportional to the logarithm of exposure (i.e., to the logarithm of the fluorescence intensity provided the exposure time is constant), the difference in optical density within the same domain in the first and in the second photographs is directly related to the ratio of fluorescence intensities observed in that domain after a 435- and 490-nm excitation.

4. pH Determination

The differences in rectified densities between the same domain of the cell in the first and the second pictures were confronted with the standard pH curve and turned into local pH values of that domain.

5. pH Maps

The local pH values were displayed either in color, resembling the traditional paper-indicator scale (red on the acidic side and blue on the alkaline side), or as contour lines connecting points of identical pH.

The standard pH curve employed in step 4 was determined as described in Chapter 4A. Small petri dishes were filled with fluorescein solutions in different buffers, then treated as a "cell sample" (i.e., two photographs were taken and their densities compared as described in Steps 2 and 3) and plotted against the pH of the buffer.

C. Technical Parameters Achieved

The pH values in the pH map are based on a two-dimensional projection of a three-dimensional object (cell). Thus, the local pH values shown represent an average pH value of a cylindrical section of that site (vertically averaged). See Figure 52.

The spatial resolution actually achieved was 150 to 200 nm (if both pictures were perfectly matched to each other). The resolution was checked by the definition of the edge of a metal particle representing a contrast object. Increasing the resolution beyond this limit by mathematical techniques (see Chapter 4D.V.G.) was also attempted.

However, as the biological object strongly disperses the fluorescence light on its way through the cell interior to the microscope objective (many reflections on various internal boundaries, variations of indices of refraction), the actual resolution inside the cell is lower (in our case, estimated to be 300 to 500 nm).

The resolution limit of the film is surely several times higher than the resolution of the optical microscope, so that the choice of the film makes no difference (even with grainy films and forced development).

The resolution of the pH map was around 0.05 to 0.10 pH units and a little more than 0.1 pH above pH 6. However, the absolute pH scale could be shifted around 0.3 pH units

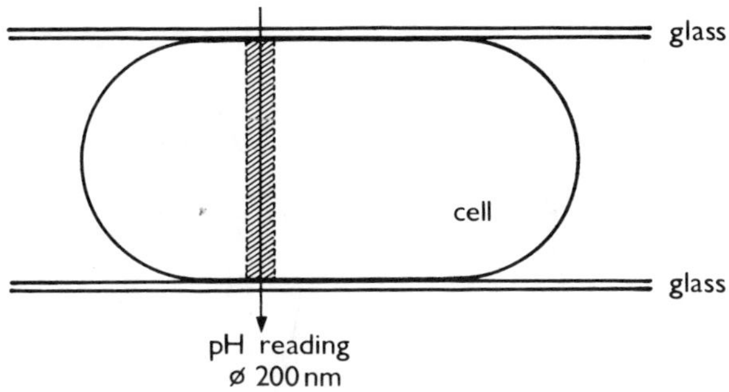

FIGURE 52. The "vertically averaged view" of the optical microscope.

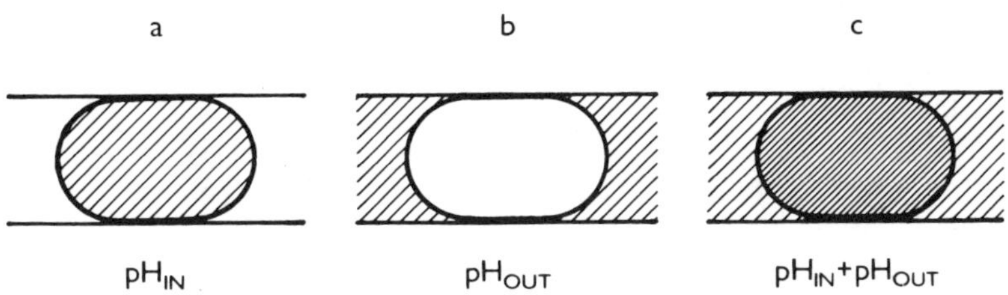

FIGURE 53. A schematic representation of the effect of incubation of cells in fluorescein diacetate (a), fluorescein (b), and fluorescein diacetate plus fluorescein (c).

in either direction due to uncertainties inherent in the densitometric procedure and in the determination of the standard pH curve.

The temporal resolution was limited by the speed of the film. In our case, the time required to take one picture was typically between 10 and 40 s. However, with an image intensifier or with a highly sensitive light detection system based on photon counting (e.g., by Hamammatsu), exposure times well below 1 s are technically feasible, even at a spatial resolution of 200 nm (higher magnification means darker pictures). So far, a temporal resolution of 9 s has been achieved using an SIT video camera[60] (measured area 1 × 1 μm) and 0.3 s using a photomultiplier[20] (spot ϕ10 μm). The latter combination gave pH readings with a standard error of the mean of about 0.04 pH units.

D. Transmembrane pH Profiles

The dye inside the cell yields information about intracellular pH, the dye outside gives information about extracellular pH values of layers close to the cell surface (Figure 53). Using a fluorescence microscope, both signals can be distinguished from each other and pH profiles across the cell membrane determined.[71]

The case (a) in Figure 53 is easy to accomplish if fluorescein diacetate is added, case (b) if fluorescein is added. If the concentration of intracellularly trapped fluorescein is at least one order of magnitude greater than the extracellular concentration of fluorescein, the extracellular signal from over- and underlying layers may be neglected and, in this way, the pH profiles across the cell membrane can be derived (case [c] in Figure 53).

E. Some Typical Results

So far, intracellular pH topography and pH profiles were studied in yeast cells[70,71] (*S. cerevisiae* and *Endomyces magnusii*), in the fungus *Penicillium cyclopium*,[72] and in the bacterium *Bacillus megaterium*.[63]

Some of the results are presented here as an example of the capabilities of the fluorescence

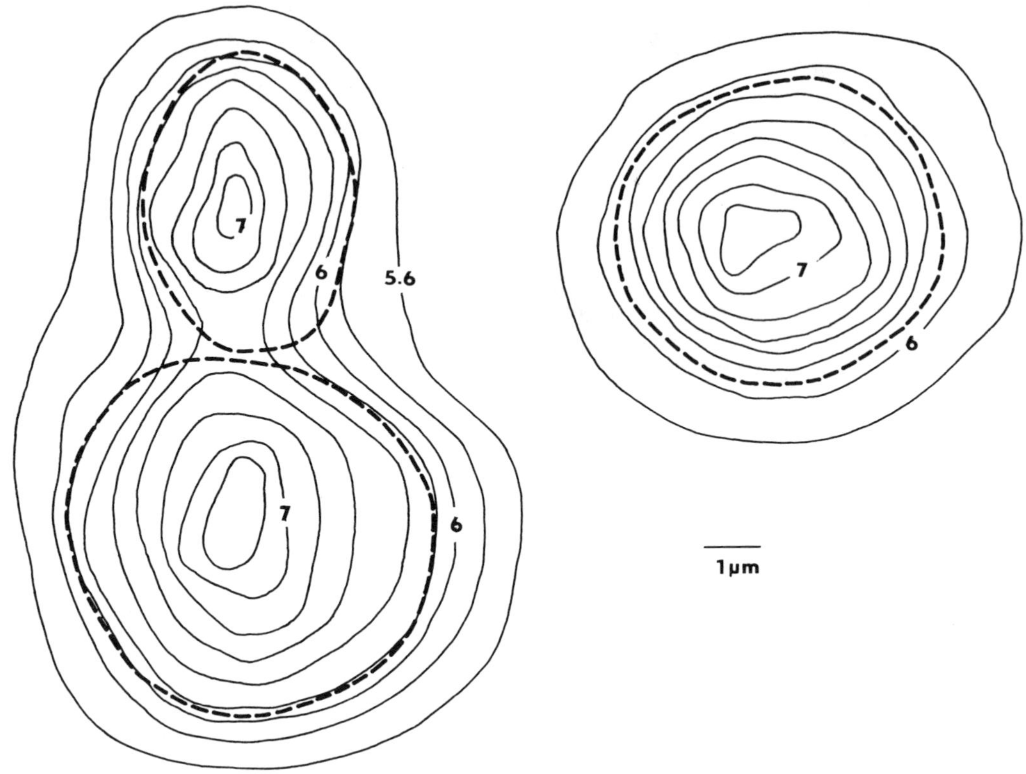

FIGURE 54. The intracellular pH map of a single and a budding yeast cell (*S. cerevisiae*). For technical reasons the maps were simplified by using pH contour lines spaced at 0.2 pH units.

pH topography technique (Figures 54, 55, and 79). For technical reasons, contour lines are presented here rather than the pseudo-color point images of cells. The discussion and interpretation of these results is presented in Chapter 6.

F. Technical Problems of Intracellular pH Topography

1. Problem 1 — Fading of Fluorescence

In optical microscopy, excitation light is focused on a tiny area (usually less than 1 mm^2). Consequently, the intensity of the exciting light is much greater than that in a cuvette of an ordinary spectrofluorometer, and, thus, significant fading of fluorescence with time may appear. (In fact, it is a gradual bleaching of the dye due to photochemical reactions.)[74]

Now there are compounds, the presence of which decreases the fading effect; second, the fluorescence intensity fades exponentially with time (so the fading may be mathematically corrected).[74] However, the safest way is to minimize the light exposure. Therefore, the use of highly sensitive films, image intensifiers, sensitive photomultipliers, or TV cameras is inevitable, especially if high spatial resolution is required. (Higher magnification means darker pictures and, thus, longer exposure times.) In our experiments, samples were observed in dim light, and full excitation intensity was used only during the actual photographic exposure of the film.

2. Problem 2 — Film

The graininess of the film, its resolution, and some fog caused by prolonged development represent no problems in intracellular pH topography.

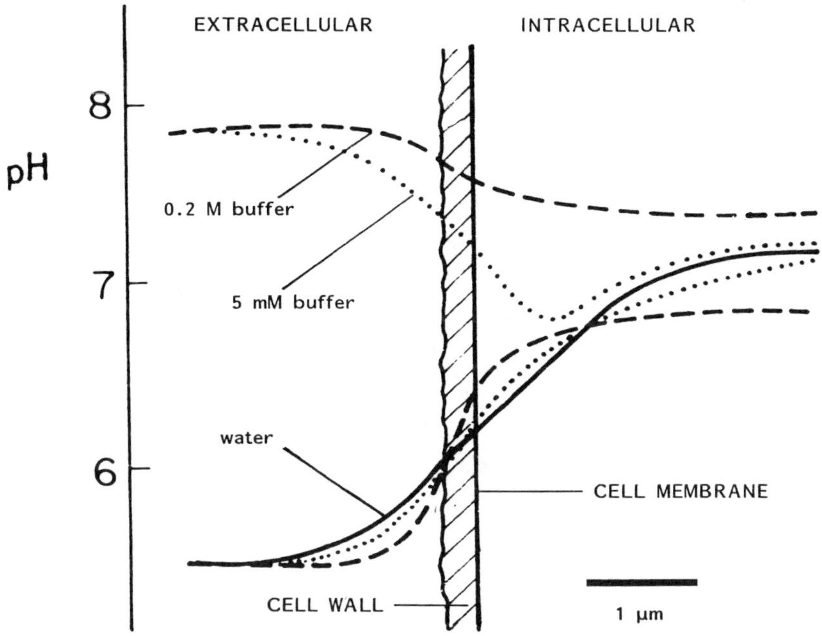

FIGURE 55. The pH profiles across the plasma membrane of the yeast *S. cerevisiae* suspended in water or buffers at pH 5 or 8.

The critical point is the nonlinearity of the densitometric curve of the film. Its shape is determined by the film (both the type of the film and the emulsion number) and the development process (Figure 56). It is preferable if the densities of both pictures are in the linear part of the S-shaped curve. This is easier to accomplish with films based on color dyes (e.g., Ilford XP® 1400, Agfapan XL® Vario-Professional®) than with classical films with the picture formed by silver grains (e.g., Ilford HP 5® or Kodak Trix®). The top-sensitivity films, such as Kodak Royal Pan®, are not recommended as their speed cannot be significantly increased by prolonged development. Usually, the "silver" films require a computerized linearization of the density curve, while the "dye" films do not (Figure 56). However, as the silver films (e.g., Ilford HP®5 with prolonged development in Microphen®) have much higher effective speeds (39 to 42 DIN or even more) than films based on color dyes (33 DIN or 1600 ASA), they are often preferred.

Both photographs, i.e., the one after 435 nm excitation and the one after 490 nm excitation, should be taken with an equal exposure time ratio. That must apply both to calibration procedures and to the determination of pH maps. The standard curve depends on the ratio of both exposure times including the Schwarzschild effect (viz., reciprocity failure, i.e., sensitivity of the film decreases if long exposure times are used). The best way to avoid these problems is to keep the exposure times of both pictures under all circumstances equal and, if possible, to diminish the brightness of the fluorescence picture excited by 490 nm with respect to that by 435 nm by inserting a gray filter with the 490-nm filter. The insertion of the gray filter is equivalent to reducing the photographic exposure and can result in shifting both pictures in a similar density area on the densitometric curve of the film. The use of a gray filter of adequate transmission value both minimizes the effect of the S-shaped densitometric curve and avoids the Schwarzschild effect. Naturally, the gray filter, once employed, must be used in all calibration procedures and in all experiments.

3. Problem 3 — TV Camera

A TV camera, if properly handled and equipped with an image intensifier, represents a

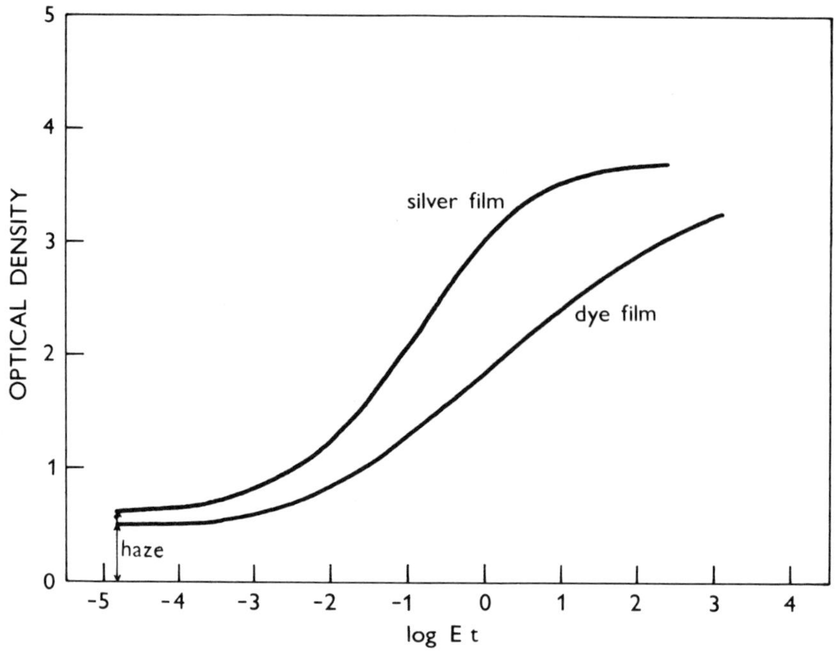

FIGURE 56. The densitometric curve of a photographic film.

far better choice than the film. However, the use of a sensitive TV camera unfortunately does not totally eliminate all above-mentioned problems concerning the nonlinearity of the film-calibration curve and of the Schwarzschild effect.

The linearity of the calibration curve of a TV camera relating the amplitude of the electrical signal and the intensity of incident light is limited by the dark current and saturation. The sensitivity may also differ both spatially and temporally. The inconvenience of the Schwarzschild effect of films is replaced by the light-memory of TV cameras. Therefore, bright light between fluorescence exposures should be avoided and dark current always substracted.

The great advantage of a TV camera lies in an instantaneous use of the experimental data (no film development) and easy storage, either in a computer or on a video cassette.[75-81]

G. Limitations of Intracellular pH Topography

The spatial resolution can be slightly improved if it is assumed that the diffraction of light distorts all images in a similar way. This distortion can be mathematically compensated by computerized image processing based on a numerical deconvolution using the following integral equation.

$$I_{obs}(x) = \int_0^{\infty} I_{true}(\xi)\, f(x - \xi) d\xi \tag{153}$$

This equation relates the fluorescence intensities in the true picture (I_{true}) with those of the observed picture (I_{obs}) by a special transfer function f of the microscope. The shape of this function can be experimentally derived (e.g., from the image of a sharp edge which substitutes for the step-like signal of Heaviside's unit step function). However, this procedure is limited by the noise of the fluorescence intensity signal, and, thus, the definition cannot be pushed too far.

A solution for the future may be in using a scanning electron beam for excitation. Hough et al.[82] reported a selective excitation of fluorescein molecules in biological samples with a resolution of several nanometers (the experimental definition limit was reported to be ~6 nm). On the other hand, the electron beam cannot selectively excite the two forms. Therefore, in such experiments a dye showing differences of fluorescence emission spectra should be employed, such as 1,4-dihydroxyphthalonitrite instead of fluorescein.

The time resolution of intracellular pH topography is limited by the sensitivity of the apparatus which probably cannot be pushed much farther, as signals from almost individual molecules are already being detected. It should be emphasized that 1 μm^3 of a 100-μM dye represents 60,000 molecules, 0.1 μm cube only 60 molecules. Thus, really individual photons arising from individual molecules are registered.

In conclusion to the discussion on the limitations of intracellular pH topography, a tempting idea, based on the shallow depth of focus of an optical microscope, would allow storage of data on different layers of the cell and, based on the principle of computerized tomography, construction of a three-dimensional pH picutre of a living cell.

REFERENCES

1. **Lakowitz, J. P.,** *Principles of Fluorescence Spectroscopy,* Plenum Press New York, 1983.
2. **Parker, C. A.,** *Photoluminescence of Solutions,* Elsevier, Amsterdam, 1968.
3. **Slavík, J.,** Anilinonaphthalene sulfonate as a probe of membrane composition and function, *Biochim. Biophys. Acta,* 694, 1, 1982.
4. Molecular Probes, Inc., *Handbook of Fluorescence Probes and Research Chemicals,* Eugene, OR, 1985.
5. **Martin, M. M. and Lindqvist, L.,** The pH dependence of fluorescein fluorescence, *J. Lumin.,* 10, 381, 1975.
6. **Leonhardt, H., Gordon, L., and Livingston, R.,** Acid-base equilibria of fluorescein and 2′,7′-dichlorofluorescein in their ground and fluorescent states, *J. Phys. Chem.,* 75, 245, 1971.
7. **Babcock, D. F. and Kramp, D. C.,** Spectral properties of fluorescein and carboxyfluorescein, *J. Biol. Chem.,* 258, 6389, 1983.
8. **Ohkuma, S. and Poole, B.,** Fluorescence probe measurement of the intralysosomal pH in living cells and the perturbation of pH by various agents, *Proc. Natl. Acad. Sci. U.S.A.,* 75, 3327, 1978.
9. **Heiple, J. M. and Taylor, D. L.,** An optical technique for measurement of intracellular pH in single living cells, in *Intracellular pH: Its Measurement, Regulation and Utilization in Cellular Functions,* Nuccitelli, R. and Deamer, D. W., Eds., Alan R. Liss, New York, 1982, 22.
10. **Weinstein, J. N., Ralston, E., Leserman, L. D., Klausner, R. D., Dragsten, P., Henkart, P., and Blumenthal, R.,** Self-quenching of carboxyfluorescein fluorescence: uses in studying liposome stability and liposome-cell interaction, in *Liposome Technology,* Gregoriadis, G., Ed., CRC Press, Boca Raton, FL, 1984, chap. 13.
11. **Rotman, D. and Papermaster, B. W.,** Membrane properties of living mammalian cells as studied by enzymatic hydrolysis of fluorogenic esters, *Proc. Natl. Acad. Sci. U.S.A.,* 55, 134, 1966.
12. **Slavík, J.,** Intracellular pH of yeast cells measured with fluorescent probes, *FEBS Lett.,* 140, 22, 1982.
13. **Thomas, J. A., Kolbeck, P. C., and Langworthy, T. A.,** Spectroscopic determination of cytoplasmic and mitochondrial pH transitions using trapped pH indicators, in *Intracellular pH: Its Measurement, Regulation and Utilization in Cellular Functions,* Nuccitelli, R. and Deamer, D. W., Eds., Alan R. Liss, New York, 1982, 105.
14. **Grinstein, S., Cohen, S., and Rothstein, A.,** Cytoplasmic pH regulation in thymic lymphocytes by an amilorde-sensitive Na^+/H^+ antiport, *J. Gen. Physiol.,* 83, 341, 1984.
15. **Horne, W. C., Norman, N. E., Schwartz, D. B., and Simons, E. R.,** Changes in cytoplasmic pH and in membrane potential of thrombin-stimulated human platelets, *J. Biochem.,* 120, 295, 1981.
16. **Piwnica-Worms, D., Jacob, R., Horres, C. R., and Lieberman, M.,** Na/H exchange in cultured chicken heart cells. pH_i regulation, *J. Gen. Physiol.,* 85, 43, 1985.
17. **Graber, M. L., Dixon, T. E., Coachman, D., Herring, A. R., Gardner, T., and Pastoriza-Munoz, E.,** Fluorescence identifies an alkaline cell in turtle urinary bladder, *Am. J. Physiol.,* 250, F159, 1986.
18. **Graber, M. L., DiLillo, D. C., Bradford, F. L., and Pastoriza-Munoz, E.,** Characteristics of fluoroprobes for measuring intracellular pH, *Anal. Biochem.,* 156, 202, 1986.
19. **Paradiso, A. M., Tsien, R. Y., and Machen, T. E.,** Na^+-H^+ exchange in gastric glands as measured with a cytoplasmic-trapped, fluorescent pH indicator, *Proc. Natl. Acad. Sci. U.S.A.,* 81, 7436, 1984.

20. **Paradiso, A. M., Tsien, R. Y., and Machen, T. E.,** Digital image processing of intracellular pH in gastric oxyntic and chief cells, *Nature,* 325, 447, 1987.
21. **Paradiso, A. M., Negulescu, P. A., and Machen, T. E.,** Na^+-H^+ and Cl^--OH^- (HCO_3^-)exchange in gastric glands, *Am. J. Physiol.,* 25, G524, 1986.
22. **Rink, T. J., Tsien, R. Y., and Pozzan, T.,** Cytoplasmic pH and free Mg^{2+} in lymphocytes, *J. Cell Biol.,* 95, 189, 1982.
23. **Chaillet, J. R. and Boron, W. F.,** Intracellular calibration of a pH sensitive dye in isolated, perfused salamander proximal tubules, *J. Gen. Physiol.,* 86, 765, 1985.
24. **Simmons, E. R., Schwartz, D. B., and Norman, N. E.,** Stimulus response coupling in human platelets: thrombin-induced changes in pH_i, in *Intracellular pH: Its Measurement, Regulation and Utilization in Cellular Functions,* Nuccitelli, R. and Deamer, D. W., Eds., Alan R. Liss, New York, 1982, 463.
25. **Schechter, E., Letellier, L., and Simons, E. R.,** Fluorescence dye as monitor of internal pH in *Escherichia coli* cells, *FEBS Lett.,* 139, 121, 1982.
26. **Rotman, A. and Heldman, J.,** Measurement of the intracellular pH of blood platelets using a photolabel fluorescent probe, *Biochim. Biophys. Acta,* 720, 75, 1982.
27. **Kano, K. and Fendler, J. H.,** Pyranine as a sensitive pH probe for liposome interiors and surfaces. pH gradients across phospholipid vesicles, *Biochim. Biophys. Acta,* 509, 289, 1978.
28. **Clement, N. R. and Gould, J. M.,** Pyranine as a probe of internal aqueous hydrogen ion concentration in phospholipid vesicles, *Biochemistry,* 20, 1534, 1981.
29. **Damiano, E., Bassilana, M., Rigaud, J. L., and Leblanc, G.,** Use of the pH sensitive fluorescence probe pyranine to monitor internal pH changes in Escherichia coli membrane vesicles, *FEBS Lett.*, 166, 120, 1984.
30. **Grzesiek, S. and Dencher, N. A.,** Dependency of ΔpH-relaxation across vesicular membranes on the buffering power of bulk solutions and lipids, *Biophys. J.,* 50, 265, 1985.
31. **Pines, E. and Huppert, D.,** pH jump: a relaxational approach, *J. Phys. Chem.,* 87, 4471, 1983.
32. **Bell, D. H., Patterson, L. K., and Gould, J. M.,** Transmembrane pH gradients and functional heterogeneity in reconstituted vesicle systems, *Biochim. Biophys. Acta,* 752, 368, 1983.
33. **Zhujun, Z. and Seitz, W. R.,** A fluorescence sensor for quantifying pH in the range from 6.5 to 8.5, *Anal. Chim. Acta,* 160, 47, 1984.
34. **Lee, H. C.,** The voltage-sensitive Na^+/H^+ exchange in sea urchin spermatozoa flagellar membrane vesicles studied with an entrapped pH probe, *J. Biol. Chem.,* 260, 10794, 1985.
35. **Garty, H., Civan, E. D., and Civan, M. M.,** Effects of internal and external pH on amiloride-blockable Na^+ transport across toad urinary bladder vesicles, *J. Membr. Biol.,* 87, 67, 1985.
36. **Ellens, H., Bentz, J., and Szoka, F. C.,** pH-induced destabilization of phosphatidyl-ethanolamine-containing liposomes, *Biochemistry,* 23, 1532, 1984.
37. **Smolarsky, M., Teitelbaum, D., Sela, M., and Gitler, C.,** A simple fluorescent method to determine complement-mediated liposome immune lysis, *J. Immunol. Methods,* 15, 255, 1977.
38. **Smith, K. K.,** Picosecond proton ejection: An ultrafast pH jump, *Chem. Phys. Lett.,* 64, 522, 1979.
39. **Chen, R. F.,** Fluorescent pH indicator. Spectral changes of 4-methylumbelliferone, *Anal. Lett.,* 1, 423, 1968.
40. **Nakashima, M., Sousa, J. A., and Clapp, R. C.,** Spectroscopic species of 4-methylumbelliferone in water and ethanol, *Nature (London),* 235, 16, 1972.
41. **Delheimer, S. C., Anderson, R. E., and Sundt, T. M., Jr.,** The measurement of spinal cord tissue pH using a diffusible, lipid-soluble, pH-sensitive fluorescent indicator, *J. Neurochem.,* 34, 1514, 1980.
42. **Gerson, D. F.,** Determination of intracellular pH changes in lymphocytes with 4-methylumbelliferone by flow cytometry, in *Intracellular pH: Its Measurement, Regulation and Utilization in Cellular Functions,* Nuccitelli, R. and Deamer, D. W., Eds., Alan R. Liss, New York, 1982, 125.
43. **Gerson, D. F. and Kiefer, H.,** High intracellular pH accompanies mitotic activity in murine lymphocytes, *J. Cell Physiol.,* 112, 1, 1982.
44. **Gerson, D. F. and Burton, A. C.,** The relation of cycling of intracellular pH to mitosis in the acellular slime mould Physarum polycephalum, *J. Cell Physiol.,* 91, 297, 1977.
45. *Biochemical and Organic Compounds for Research and Diagnositc Clinical Reagents,* Sigma Chemical Co., St. Louis, MO, 1986.
46. **Schummer, U. and Schiefer, H. G.,** Surface potential of mycoplasma membranes, *FEMS Microbiol. Lett.,* 23, 143, 1984.
47. **Alpes, H. and Pohl, W. G.,** 3-palmitoyl-7-oxycoumarin: a new lipoid pH indicator, *Naturwissenschaften,* 65, 652, 1978.
48. **Valet, G., Raffael, A., Moroden, L., Wünsch, E., and Ruhenstroth-Bauer, G.,** Fast intracellular pH determination in single cells by flow cytometry, *Naturwissenschaften,* 68, 265, 1981.
49. **Kurz, I. and Balaban, R. S.,** Fluorescence emission spectroscopy of 1,4-dihydroxyphthalonitrile, *Biophys. J.,* 48, 499, 1985.

50. **Lee, H. C. and Forte, J. G.,** A novel method for measurement of intravesicular pH using fluorescent probes, *Biochim. Biophys. Acta,* 601, 152, 1980.
51. **Rogers, J., Hesketh, T. R., Smith, G. A., Beaven, M. A., Metcalfe, J. C., Johnson, P., and Garland, P. B.,** Intracellular pH and free calcium changes in single cells using quene 1 and quin 2 probes and fluorescence microscopy, *FEBS Lett.,* 161, 21, 1983.
52. **Rogers, J., Hesketh, T. R., Smith, G. A., and Metcalfe, J. C.,** Intracellular pH of stimulated thymocytes measured with a new fluorescent indicator, *J. Biol. Chem.,* 258, 5994, 1983.
53. **Hesketh, T. R., Moore, J. P., Morris, J. D. H., Taylor, M. V., Rogers, J., Smith, G. A., and Metcalfe, J. C.,** A common sequence of calcium and pH signals in the mitogenic stimulation of eukaryotic cells, *Nature,* 313, 481, 1985.
54. **Bashford, C. L.,** Spectrophotometry and fluorimetry of cellular compartments, in *Spectrophotometry and Spectrofluorimetry — A Practical Approach,* Harris, D. A., and Bashford, C. L., Eds., IRL Press, Oxford, 1987, chap. 5.
55. **Lee, H. C., Forte, J. G., and Epel, D.,** The use of fluorescent amines for the measurement of pH_i: applications in liposomes, gastric microsomes and sea urchin gametes, in *Intracellular pH: Its Measurement, Regulation and Utilization in Cellular Functions,* Nuccitelli, R. and Deamer, D. W., Eds., Alan R. Liss, New York, 1982, 135.
56. **Reynolds, G. T.,** Image intensification applied to microscope systems, *Adv. Opt. Electron Microsc.,* 2, 1, 1968.
57. **Reynolds, G. T.,** Image intensification applied to biological problems, *Q. Rev. Biophys.,* 5, 295, 1972.
58. **Reynolds, G. T. and Taylor, D. L.,** Image intensification applied to light microscopy, *Bioscience,* 30, 586, 1980.
59. **Rost, F. W. D.,** A microspectrofluorometer for measuring spectra of excitation, emission and absorption in cells and tissues, in *Fluorescence Techniques in Cell Biology,* Thaer, A. and Sernetz, M., Eds., Springer-Verlag, Berlin, 1973, 57.
60. **Tanasugarn, L., McNeil, P., Reynolds, G. T., and Taylor, D. L.,** Microspectrofluorometry by digital image processing: measurement of cytoplasmic pH, *J. Cell Biol.,* 98, 717, 1984.
61. **Eiger, M.,** Proton transfer acid-base catalysis and enzymatic hydrolysis. I. Elementary processes, *Angew. Chem. Int. Ed. Engl.,* 3, 1, 1964.
62. **Selvaggio, A. M., Schwartz, J. H., Bengele, H. H., and Alexander, E. A.,** Kinetics of the Na^+-H^+ antiporter as assessed by the change in intracellular pH in MDCK cells, *Am. J. Physiol.,* 251, C558, 1986.
63. **Slavík, J.,** unpublished data.
64. **Hadam, M. R.,** Flow cytometry and surface-marker phenotyping using monoclonal antibodies: a combined approach to precisely define the state of the immune system, in *New Aspects in Physiological Antitumor Substances,* S. Karger, Basel, 1985, 120.
65. **Beverly, P. C. L.,** The use of the fluorescence-activated cell sorter for the identification and analysis of function of cell subpopulations, in *Monoclonal Antibodies in Clinical Medicine,* McMichael and Fabre, Eds., Academic Press, New York, 1982, 557.
66. **Kruth, H. S.,** Flow cytometry: rapid biochemical analysis of single cells, *Anal. Biochem.,* 125, 225, 1982.
67. **Melaned, M. R., Mullaney, P. F., and Mendelsohn, M. L., Eds.,** *Flow Cytometry and Cell Sorting,* John Wiley & Sons, New York, 1979.
68. **Visser, J. W. M., Jongeling, A. A. M., and Tanke, H. J.,** Intracellular pH-determination by fluorescence measurements, *J. Histochem. Cytochem.,* 27, 32, 1979.
69. **Gerson, D. F. and Kiefer, H.,** Intracellular pH and the cell cycle of mitogen-stimulated murine lymphocytes, *J. Cell. Physiol.,* 114, 132, 1983.
70. **Slavík, J.,** Intracellular pH topography: determination by a fluorescent probe, *FEBS Lett.,* 156, 227, 1983.
71. **Slavík, J. and Kotyk, A.,** Intracellular pH distribution and transmembrane pH profile of yeast cells, *Biochim. Biophys. Acta,* 766, 679, 1984.
72. **Roos, W. and Slavík, J.,** Intracellular pH topography of Penicillium cyclopium protoplasts. Maintenance of ΔpH by both passive and active mechanisms, *Biochim. Biophys. Acta,* 899, 67, 1987.
73. **Taylor, D, L., Waggoner, A. S., Larri, F., Murphy, R. F. and Birge, R. R.,** Applications of Fluorescence in the Biomedical Sciences, Alan R. Liss, New York, 1986.
74. **Giloh, H. and Sedat, J. W.,** Fluorescence microscopy: Reduced photobleaching of rhodamine and fluorescein protein conjugates by n-propyl gallate, *Science,* 217, 1252, 1982.
75. **Rich, E. S. and Wampler, J. E.,** A flexible, computer-controlled video microscope capable of quantitative, spatial, temporal, and spectral measurements, *Clin. Chem,* 27, 1558, 1981.
76. **Inoué, S.,** Video-image processing greatly enhances contrast, quality and speed in polarization-based microscopy, *J. Cell Biol.,* 89, 346, 1981.
77. **Kudo, Y., Ozaki, K., Miyakawa, A., Amano, T., and Ogura, A.,** Monitoring of intracellular Ca^{2+} elevation in a single neural cell using a fluorescence microscopy/video-camera system, *Jpn. J. Pharmacol.,* 41, 345, 1986.

78. **Kudo, Y., Ito, K., Miyakawa, H., Izumi, Y., Ogura, A., and Kato, H.,** Cytoplasmic calcium elevation in hippocampal granule cell-induced by perforant path stimulation and L-glutamate application, *Brain Res.*, 407, 168, 1987.
79. **Allen, R. D.,** Video-enhanced microscopy, in *Optical Methods in Cell Physiology,* Deweer, P. and Salzberg, B. M., Eds., John Wiley & Sons, New York, 1986, 3.
80. **Ellis, G. W., Inoué, S., and Inoué, T.,** Computer-aided light microscopy, in *Optical Methods in Cell Physiology,* Deweer, P. and Salzberg, B. M., Eds., John Wiley & Sons, New York, 1986, 15.
81. **Fay, F. S., Fogarty, K. E., and Coggins, J. M.,** Analysis of molecular distribution in single cells using a digital imaging microscope, in *Optical Methods in Cell Physiology,* Deweer, P. and Salzberg, B. M., Eds., John Wiley & Sons, New York, 1986, 51.
82. **Hough, P. V. C., McKinney, W. R., Ledbetter, M. C., Pollack, R. E., and Moos, H. W.,** Identification of biological molecules in situ at high resolution via the fluorescence excited by a scanning electron beam, *Proc. Natl. Acad. Sci. U.S.A.*, 73, 317, 1976.

Chapter 4E

ASSAYS OF INTRACELLULAR pH USING CHEMICAL PROBES: NUCLEAR MAGNETIC RESONANCE SPECTROSCOPY

Jan Slavík

TABLE OF CONTENTS

I. THEORETICAL ASPECTS

A. Introduction

Nuclear magnetic resonance (NMR) spectroscopy passed through an era of rapid development. This relatively young technique, excitingly young if compared with optical absorption or fluorescence methods, was introduced into physical and chemical research in 1946. The earliest reports on biological objects are probably those on experiments with blood and blood cells, performed by Odeblad et al.[1] and by Jardetzky and Wertz[2] in 1956.

High-resolution NMR gained wide recognition as a powerful analytical technique in biological and medical research only after the introduction of the technical improvements of the late 1960s (pulsed Fourier transformation NMR and superconducting magnets), which dramatically enhanced its sensitivity. This opened the way for NMR into biology and medicine and resulted in a burst of papers in the 1970s. Among these pioneering papers was that by Moon and Richards[3] who estimated the intracellular pH value in red blood cells using the chemical shift of ^{31}P-phosphate magnetic resonance in 1973 and, thus, laid the cornerstone for NMR intracellular measurements.

In the 1980s NMR techniques have become very popular. The access to NMR instruments is now easier, the handling simpler, and the number of workers acquainted with the technique is growing rapidly. The recent development of NMR tomography offers the possibility of two-dimensional or even three-dimensional NMR imaging, rendering information on the local states of metabolism (including pH values) at different sites inside the body or tissue.

The advantage of the NMR technique lies in its truly noninvasive and nondestructive nature. A relatively simple measurement yields a large amount of information.

The principal disadvantage of NMR lies in its inherently low sensitivity, requiring a high density of the biological material and/or long times of NMR-signal acquisition which potentially may invade and destroy the biological sample. NMR spectra of a biological sample are often complicated and difficult to interpret, even for an expert. The mere recording of an NMR spectrum requires a great amount of skill and considerable know-how. Practice has shown that multidisciplinary groups had to be formed to run NMR spectrometers dedicated to biological measurements in order to cope with both the use of the spectrometer and the demands imposed by the living system.

In the case of intracellular pH measurements, it should be carefully considered whether such a complex technique as NMR is necessary to answer the relatively simple question of what the value of intracellular pH is. Regarding the time and financial expense at the present state of the art of the technique, NMR should be used preferentially for studying complex and dynamic phenomena in cellular and vesicular systems rather than for simple intracellular pH measurements.

We are aware of the fact that textbooks on NMR spectroscopy can be obtained everywhere in the world and that it is beyond the scope of this book to give a comprehensive, though concise, general introduction to NMR spectroscopy. Bearing in mind the above-mentioned existence of the multidisciplinary groups, the following text covers only rather oversimplified basic information on NMR theory and experiment, indispensable for understanding the principles of intracellular pH experiments. However, all the specialized aspects of the pH measurement are discussed in detail. Special care was given to the questions of reproducibility, calibration, accuracy of pH_{in} data, and to the summary of properties of particular chemical substances used for pH measurement. The theory of NMR methods and relevant experimental techniques are discussed in detail in monographs.[4-9] Much practical information on NMR experiments is found in the book written by Fukushima and Roeder.[10]

A clear and concise introduction to intracellular pH measurement can be found in specialized review articles, especially by Roberts and Jardetzky[11] (monitoring of cellular metabolism by NMR), Gordon[12] (medicine and NMR), Jacobson and Cohen[13] (intracellular pH measurement by NMR), and Williams and Gadian[14] (tissue metabolism studies *in vivo* by NMR).

B. Atomic Nucleus

Virtually every element comprises an isotope with a nucleus possessing a non-zero spin. These nuclei behave like tiny bar magnets. The easiest way to study their behavior is to place them in a strong, homogeneous, steady-state magnetic field. The magnetic field makes the magnetic properties of the nuclei observable by means of another magnetic field. This latter, observing field is used for ''communicating'' with the nucleus. The signal from the nucleus appears as an attenuation or enhancement of the observing field in continuous wave techniques or as a response following a short, strong observing pulse in pulse techniques (cf. Chapter 4E. III. C).

From this standpoint, the nucleus can be considered as a tiny submolecular reporter that passes information about the structure of the particular molecule it is part of and about other molecules and their movements in its immediate neighborthood. This information can be derived from the changes of the NMR parameters. Therefore, if a suitable chemical substance undergoing a pH-dependent transition is chosen, it can be used as an indicator of the local pH value. For instance, the transition $H_2PO_4^- \leftrightharpoons HPO_4^{2-}$ with pK ~6.9 can be followed by ^{31}P NMR, and, thus, the H^+ concentration around these phosphate molecules can be measured.

C. The Phenomenon of NMR

The behavior of a nuclear spin in a homogeneous steady-state magnetic field can be described either classically or quantum-mechanically (Figure 57).

The classical description is more illustrative. It depicts the nucleus as a tiny bar magnet spinning either parallel or antiparallel to the direction of the external magnetic field, like a gyroscope in a gravitational field. The precession frequency is called Larmor frequency. If the frequency of the observing radio-frequency field exactly matches the rotational frequency of a parallel spin (the resonance effect), the spin in question flips over and continues to rotate antiparallel. The relaxation processes (e.g., due to interaction with other spins) eventually cause the spin to flip back (relax) to parallel spinning.

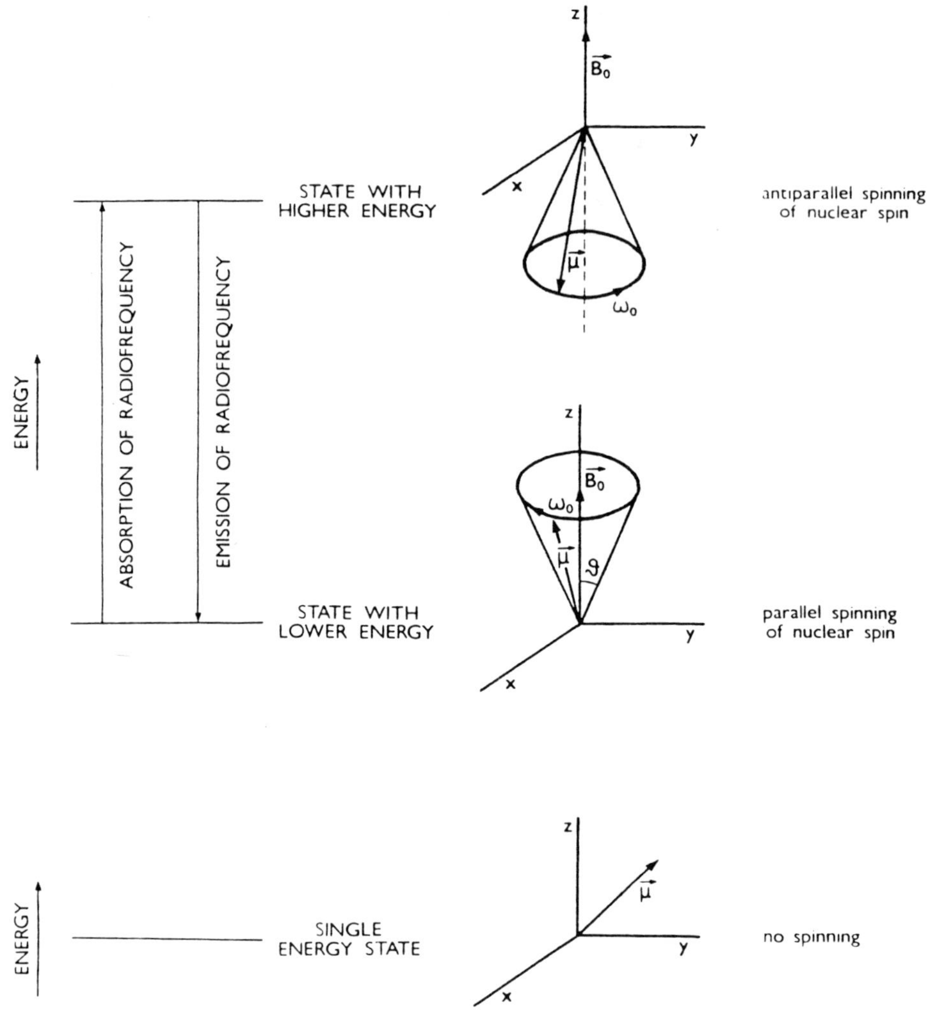

FIGURE 57. The splitting of energy of nuclei and the precession of the nuclear spin induced by switching on magnetic field B_0 (for spin number = 1/2).

The quantum-mechanical approach is based on a description of the energy of the nucleus. In a magnetic field each nucleus can exist in two states with different energy (the simplest case with spin number 1/2). The small difference in energy arises from the interaction of the nuclear spin with the magnetic field. The particular nucleus can be either in the state with a lower energy or in the state with a higher energy. If the energy corresponding to the difference is supplied by the observing electromagnetic field, the nucleus can undergo transition from the former state to the latter. The transition is called nuclear magnetic resonance as it takes place if, and only if, the energy supplied exactly equals the energy difference between the two states. The difference between the levels is given by the following quantum-mechanical formula:

$$\Delta E = \frac{\gamma h}{2\pi} B_0 \qquad (154)$$

and the second (observing) electromagnetic field must have the frequency ω fulfilling the condition:

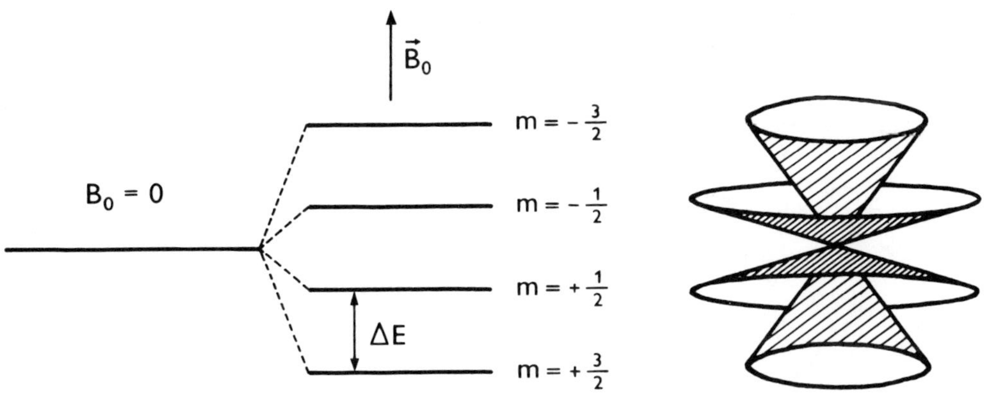

FIGURE 58. The quadruple splitting of energy and four possible modes of spin precession for nuclei with spin number = 3/2.

$$\Delta E = h\omega/2\pi \tag{155}$$

if NMR is to appear (where γ is the gyromagnetic ratio, h is Planck's constant, and B_0 the intensity of the magnetic field, i.e., magnetic induction).

In the case of isotopes with nuclear spin numbers greater than 1/2 (e.g., ^{35}Cl), the nucleus in the magnetic field can exist in more than two energy states (^{35}Cl in four) or, according to the classical description, there are more than two orientations of the spin relative to the direction of the magnetic field (Figure 58).

D. What Produces the NMR Signal?

In theory, any nucleus can be observed by NMR spectroscopy (nuclei with zero spins via their quadrupole moments).

In practice, only a few nuclei (all of them with nonzero spins) are employed in biological and medical research, and the majority of experiments are mostly limited to ^{1}H, ^{13}C, ^{15}N, ^{19}F, and ^{31}P. The NMR properties of these and other isotopes of biologically important elements are summarized in Table 6.

E. What Is Measured?

The basic NMR parameters used for characterization of a particular chemical group are the following:

- *Intensity* — intensity of the radio-frequency signal absorbed or emitted by the nuclei (either the amplitude or the peak area in Figure 59)
- *Line width* — the width of the peak at half-maximum intensity of the signal
- *Chemical shift* — the shift in resonance frequency of the investigated nuclei (ν) relative to resonance frequency of the reference signal (ν_{ref}), i.e., separation of peaks between the reference nucleus and the investigated nucleus (Defined as

$$\delta = \frac{\nu - \nu_{ref}}{\nu_{ref}} 10^6 \tag{156}$$

  it is independent of B_0. It is expressed in parts per million [ppm].)
- *Longitudinal relaxation* time T_1 (also called spin-lattice relaxation time) and transversal relaxation time T_2 (also called spin-lattice relaxation time) — used for the characterization of relaxation processes
- *Coupling constant J* — a difference in resonance frequency between two neighboring peaks of a multiplet

Table 6
PROPERTIES OF SELECTED NUCLIDES

Nuclide	Nuclear spin quantum number	Resonance frequency for B = 10T (MHz)	Sensitivity relative to ^{31}P (relative U)	Natural isotopic abundance (%)	Receptivity[a] (relative U)
^{1}H	1/2	425.8	15.1	99.984	15.6
^{2}H	1	65.3	0.14	0.016	0.0002
^{13}C	1/2	107.0	0.24	1.1	0.0026
^{14}N	1	30.8	0.015	99.6	0.0151
^{15}N	1/2	43.1	0.016	0.36	0.00006
^{17}O	5/2	57.7	0.44	0.004	0.00016
^{19}F	1/2	400.6	12.6	100	12.54
^{23}Na	3/2	112.6	1.4	100	1.39
^{25}Mg	5/2	26.1	0.04	10.0	0.0041
^{29}Si	1/2	84.6	0.12	4.7	0.0055
^{31}P	1/2	172.4	1	100	1
^{34}S	3/2	32.7	0.03	0.74	0.00025
^{35}Cl	3/2	41.7	0.07	75.4	0.053
^{37}Cl	3/2	34.7	0.04	24.6	0.01
^{39}K	3/2	19.9	0.008	93.1	0.0071
^{41}K	3/2	10.9	0.001	6.9	0.00008
^{43}Ca	7/2	28.6	0.096	0.13	0.00013
^{51}V	7/2	111.9	5.7	100	5.74
^{55}Mg	5/2	105.0	2.6	100	2.64
^{127}I	5/2	85.2	1.4	100	1.41

[a] Sensitivity relative to ^{31}P with respect to natural isotopic abundance.

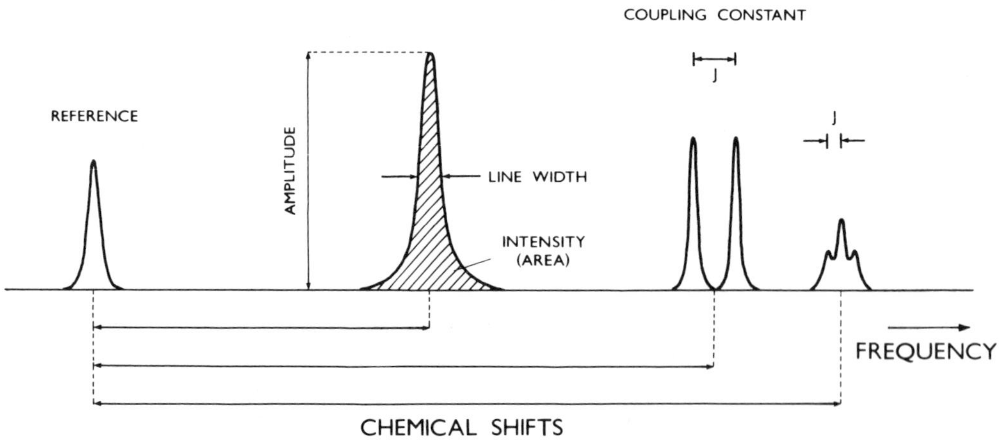

FIGURE 59. A schematic illustration of basic parameters derived from a high-resolution NMR spectrum.

F. Basic Relations Between NMR Parameters

Intensity of the signal (more precisely, the area under the peak) — This is directly proportional to the number of nuclei present in the sample. A law analogous to the Lambert-Beer Law of optical spectroscopy is valid, if, like in optics, saturation is avoided. Therefore, relative concentrations of chemical groups can be estimated and conclusions about chemical structure drawn.

The NMR spectrum — This generally consists of a large number of peaks. The higher

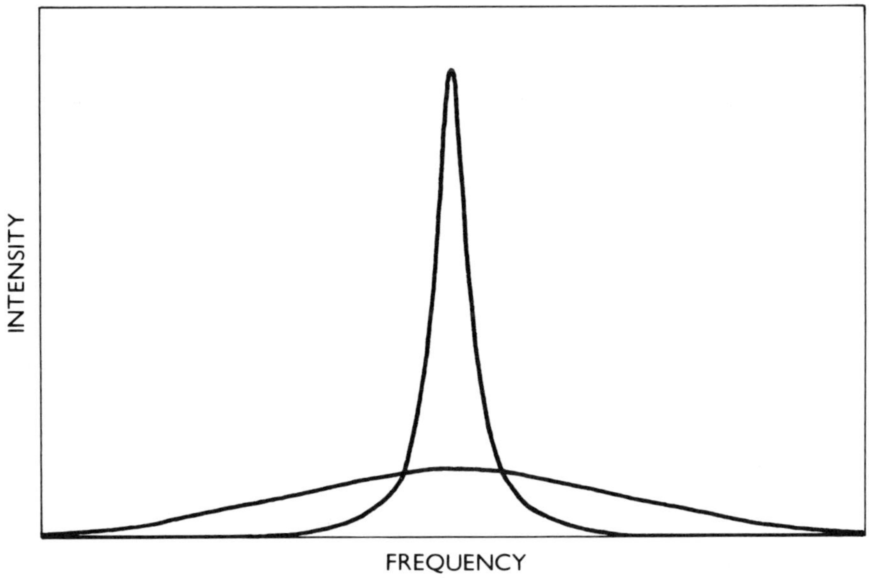

FIGURE 60. Broadening of a resonance peak corresponding to a tenfold decrease of the relaxation time T_2.

the intensity of the steady magnetic field B_0, the better the resolution (wider spacing) of the peaks. The width of the peak and its shape are related to its relaxation time T_2. A decrease in T_2 (e.g., due to the binding or due to some restriction of motion of that group) is accompanied by an appropriate broadening of the peak, or

$$\text{half-width} \cong 1/T_2 \tag{157}$$

This results in a lower height of the peak and, consequently, the broadened peak may disappear due to background noise (Figure 60).

Chemical shift — The actual resonance frequency of a particular nucleus is given by the isotope, by the type of bond of the atom, and by the structure of the molecule (see Figure 61). The interactions influencing the resonance frequency are extremely complex and allow virtually any chemical group to be distinguished by its chemical shift value. Consequently, NMR spectra of biological samples look rather complicated (Figures 61D and 61E). The values of chemical shifts are always related to a reference compound (phosphoric acid in Figure 61).

The reference compound — This may be any compound either present in the measured sample (so-called internal reference) or separated from the sample by a diffusion barrier, such as in a thin capillary inside the NMR tube (so-called external reference). The reference peak should be sharp and its resonance frequency stable. If used for intracellular pH measurement, the peak of such a standard should be truly pH independent.

The relaxation processes — These are responsible for the return of the nucleus from the state of higher energy back to the state of lower energy. They are caused, e.g., by direct interactions of spins of neighboring nuclei or by the interaction of nuclear spin with fluctuating magnetic fields in turn caused by the movement of neighboring molecules. Thus, any molecular or intramolecular motion may be reflected in NMR if it occurs on the time scale close to the Larmor precession frequency of the nuclear spin (translation, rotation, or reorientation of whole molecules or groups within a molecule). The relaxation processes in NMR resemble the vibrational-rotational relaxations in optics, but they are much slower

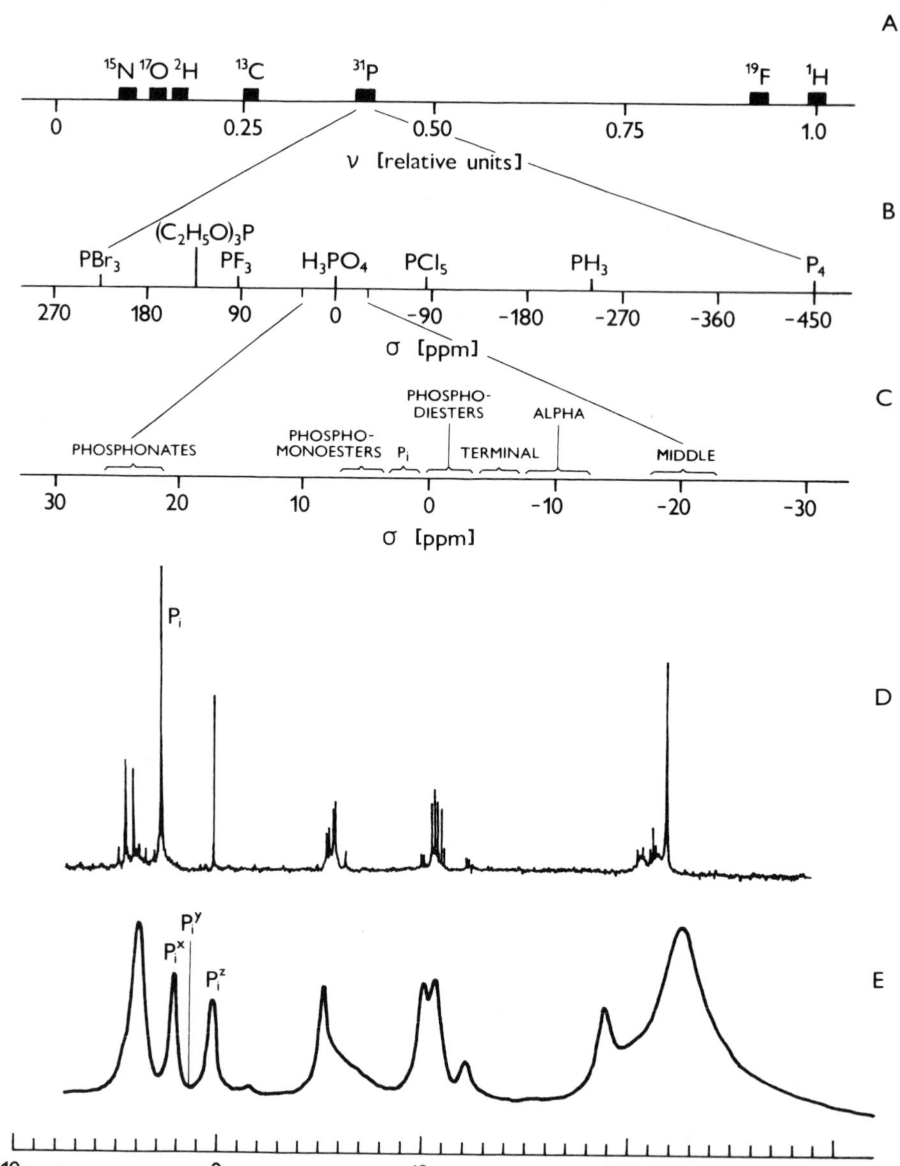

FIGURE 61. Schematic differentiation of resonance frequencies with respect to the isotope (A), the type of bonds (B), and the type of chemical compound (C). Comparison of ^{31}P NMR spectra of a perchloric acid extract of cells (D) with the *in vivo* intracellular signal (E) gives an indea of the experimentally observed intracellular line broadening and of the effect of compartmentation of orthophosphate in three regions differing in pH (denoted as P_i^x, P_i^y, P_i^z), inside the yeast *Saccharomyces cerevisiae*. (Adapted from Reference 52).

(milliseconds to seconds compared to picoseconds or nanoseconds in optics). As a consequence, the slow relaxation rate limits the repetition frequency in pulse FT NMR experiments and in this way also decreases the sensitivity of the NMR technique and limits the temporal resolution of NMR experiments (cf. Chapter 4E. III. D).

The spin-spin coupling — This appears due to the interaction of neighboring nuclei and results in splitting of peaks into multiplets of peaks. As it may complicate the interpretation

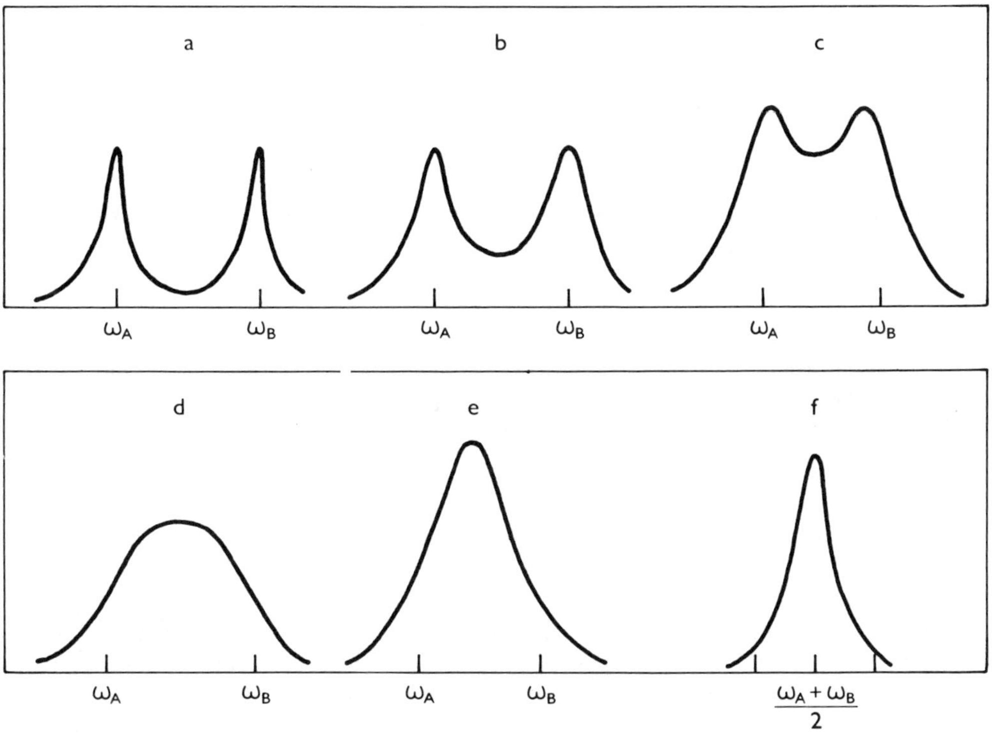

FIGURE 62. Effect of the exchange rate on the shape of the NMR spectrum of two rapidly exchanging compounds A and B. The corresponding values of $\Delta\tau(\omega_A - \omega_B)$ are approximately 100 (a), 10 (b), 5 (c), 1 (d), 0.1 (e), and 0.01 (f).

of spectra and their measurement, special spin-decoupling techniques removing this splitting are often employed.

G. Theory of pH Measurement

1. Exchange Reaction

The observation of a single resonance peak of the pH indicator instead of two or more peaks of its particular forms is a startling NMR phenomenon uncommon in absorption or fluorescence spectroscopy. In NMR, two or more different chemical species can be represented by a single NMR resonance peak provided they mutually exchange rapidly enough (i.e., the rate of exchange is higher than the frequency difference between the two resonances). A pH indicator which rapidly fluctuates between the dissociated and undissociated form exchanging protons with the medium is a typical example.

Let us consider a simple case of two interchanging species, each of them characterized by a single peak with resonance frequencies of ω_A and ω_B, respectively. Depending on the difference between their resonance frequencies $\omega_A - \omega_B$ and the rate of mutual exchange (in the absence of spin-spin coupling), the following different basic shapes of NMR spectra (Figure 62) are obtained. The slowly interchanging species represented by two single narrow peaks (Figure 62a) broaden with increasing rate of exchange (Figure 62b,c), then merge in one broad peak (Figure 62d, e) and finally appear as a single narrow peak (Figure 62 f). In this last case (f), the resonance frequency corresponding to the resulting peak is a weighted arithmetical average of both interchanging species.

Quantum theory offers an explanation based on Heisenberg's uncertainty principle, relating the uncertainty in the determination of energy to that of time:

$$\Delta E \, \Delta t \cong h/2\pi \tag{158a}$$

or equivalently expressed in frequency and time

$$\Delta E \, \Delta t \cong {}^1/_2\pi \tag{158b}$$

Let $\Delta\tau$ denote the mean time between two consecutive chemical exchanges. The broadening of spectral lines should be in the range of

$$\Delta\nu = {}^1/_2\pi\Delta\tau \tag{159}$$

For $\Delta\tau = 10^{-1}$, 10^{-3} and 10^{-5} s, we obtain line broadenings of about 1, 100, and 10^4 Hz, respectively. Obviously, such values are too small to be resolved experimentally in optical spectroscopy, but on the NMR scale they play a significant role.

2. pH-Dependent Peaks in NMR Spectrum

In contrast with absorption and fluorescence spectroscopy, the pH-indicating substance may be represented by a single peak in the NMR spectrum, the position of which is pH dependent. As explained above, this is because both forms of the pH-indicating substance equilibrate so fast and their resonance frequencies are so close to each other than conditions of appearance of one single narrow peak (Figure 62) are virtually always fulfilled. Thus, a typical difference in the chemical shift between two related forms of the pH indicator is 2 to 5 ppm, which corresponds to a difference in resonance frequencies $\Delta\omega = 400$ to 1000 Hz (at 200 MHz). If the two forms interchange at a rate of 30,000 Hz, then $\Delta\omega\Delta\tau$ is approximately equal to 0.01 to 0.03, and a single peak should, thus, appear.

The position of the peak of two rapidly exchanging species can be calculated as follows. Let ω_A and ω_B denote the values of resonance frequencies of A and B, τ_A and τ_B the average lifetime of each form, and c_A and c_B their concentrations. The resonance frequency observed is a weighted average of resonances of both forms:

$$\omega = \tau_A c_A \omega_A/(\tau_A c_A + \tau_B c_B) + \tau_B c_B \omega_B/(\tau_A c_A + \tau_B c_B) \tag{160}$$

On the assumption that the time during which the first form exists is, on the average, equal to the time during which the second form exists ($\tau_A = \tau_B$), Equation 160 simplifies to

$$\omega = c_A \omega_A/(c_A + c_B) + c_B \omega_B/(c_A + c_B) \tag{161}$$

In terms of chemical shifts,

$$\delta = c_A \delta_A/(c_A + c_B) + c_B \delta_B/(c_A + c_B) \tag{162}$$

The weighting coefficients are concentrations (not chemical activities) of the two forms, and the equation characterizing the pH-indicating behavior of the substance is (cf. Equations 80 and 81):

$$\mathrm{pH} = \mathrm{p}K' + \log(c_B/c_A) \tag{163}$$

A combination of Equations 162 and 163 yields a convenient expression

$$\mathrm{pH} = \mathrm{p}K' + \log(\delta_A - \delta)/(\delta - \delta_B) \tag{164}$$

often used in the interpretaion of NMR spectra.

However, the validity of this relation is rather limited, as *in vivo* there are usually more than two interchanging forms of a pH indicator which contribute to the spectrum. For instance, the ^{31}P NMR signal of phosphate-containing pH indicators consists also of contributions by various complexes of the pH-indicating substance with divalent cations, basic amino acids, and peptides. These contributions cannot be neglected in the interpretation of spectra (see the section on calibration). Therefore, the more general case of several chemical species contributing to the measured peak must be considered. Equation 162 is then replaced with the weighted arithmetical average of all components with respect to their relative concentrations:

$$\delta = \sum_{j} c_j\delta_j/c \tag{165}$$

where c denotes the total concentration ($c = \Sigma_j c_j$), provided that all "lifetimes" are equal to each other: $\tau_1 = \tau_2 = \tau_3 \ldots$ Unfortunately, Equation 165 cannot be transformed into an equation similar to Equation 164 as each of the components has a specific pK' value.

II. pH INDICATORS FOR NMR

A. Introduction

Before starting to measure intracellular pH by NMR, it should be clear (1) which nuclide and (2) which metabolite will be used.

1. Nuclide

Nuclei with high receptivity should be preferred as sensitivity is the weak point of NMR spectroscopy. It is seen in Table 6 that ^{1}H, ^{13}C, ^{14}N, ^{15}N, ^{17}O, ^{19}F, and ^{31}P may be convenient for biological measurement. Indeed, many of them, viz., ^{1}H, ^{15}N, ^{19}F, and ^{31}P, have already been employed for intracellular pH measurement. Among these isotopes, ^{31}P appears to be the most attractive as ^{31}P NMR spectra simultaneously offer data both about pH and about the physiological state of the cell under study. The ^{31}P experiments do not require feeding with an isotopically labeled compound and, in contrast to ^{1}H-NMR, the simple one-pulse FT NMR can be used to obtain spectra of satisfactory quality. In addition, ^{31}P spectra are usually simpler than their ^{13}C and ^{1}H counterparts and, hence, readily interpretable.

The following review shows the principal trends of NMR intracellular pH measurement, but does not represent an exhaustive review of all the substances used in all papers dealing with NMR pH measurement.

2. Metabolite

The metabolites generally in abundance in the cell are preferred because of sensitivity limitations. It is also convenient if the chosen compound is generally present in most types of cells, and gives a well-distinguished, easily interpretable peak (or peaks), and has a pK near the middle of the pH range measured. These reasons have led to the wide use of inorganic phosphate ^{31}P NMR for intracellular pH measurement though, from the pH-indication point of view, it is far from ideal.

The reference compound should have a sharp, pH-independent peak not interfering with the spectral position of the pH-indicating compound. However, as any peak may, in principle, serve as a reference, a proper combination of two pH-indicating compounds could compensate for the undesirable pK shifts induced by ionic strength if both compounds have a similar pK and the pH-induced change of the chemical shift has an equal magnitude but the opposite sign (e.g., inorganic phosphate relative to methyl phosphonate). Also, the other peak of the same compound may be used as a reference (e.g., the difference between chemical shifts of alpha- and gamma-phosphate of ATP in ^{31}P NMR or the shift between the two central form peaks of difluorinated alanine in ^{19}F NMR).

B. ^{1}H NMR

Theoretically, ^{1}H NMR should be the most convenient technique. The ^{1}H nucleus possesses the highest sensitivity in comparison with other nuclei (Table 6), and almost every compound is capable of yielding an ^{1}H NMR signal.

In practice, however, the ^{1}H NMR spectrum of a biological sample is dominated by the signal of cellular water (pure water represents 110 M hydrogen ions) and often by endogenous lipids. The distinction of a signal from a 1 m*M* concentration of the metabolite of interest, from the large and uninteresting background, requires the use of elaborate techniques. This approach may include not only special spin-echo pulse-sequence techniques and various "editing" methods,[11,16] but also the replacement of H_2O molecules with D_2O. It is clear that a direct measurement of pH via observation of a signal arising from a 0.1 μ*M* concentration of free proton corresponding to pH 7 is pure fantasy. Furthermore, the small chemical shift range of biological compounds (10 ppm in ^{1}H NMR compared to 40 ppm in ^{31}P and 200 ppm in ^{13}C), results in numerous overlaps of peaks and severely limits the practical use of ^{1}H NMR as many more compounds contribute to ^{1}H NMR than to ^{31}P NMR spectra.

Nevertheless, using the above techniques, intracellular ^{1}H NMR resonances of some compounds can be distinguished *in vivo*, e.g., phosphocholine, phosphocreatine, creatine, aspartate, glutamate, alanine, lactate, and others.[17] Based on these results, intracellular pH was estimated in intact erythrocytes by detecting the C_2 proton resonance of a histidyl residue of hemoglobin[18] and the imidazole resonance (C_2 proton and C_4 and C_5 protons) of hemoglobin.[19] Quite recently, anserine[16] and carnosine[20,21] were employed for intracellular-pH measurements of intact muscle.

The reference compound is usually external tetramethylsilane (TMS) or sodium dimethylsilapentasulfonate (DSS).

C. ^{13}C NMR

The ^{13}C NMR signal is quite sensitive to its chemical environment, and there is a strong interaction with protons. Thus, in principle, any compound with a suitable p*K* value can be employed for intracellular pH measurements. The use of amino acids would be convenient, and there is a temptation to use bicarbonate.

However, ^{13}C is naturally rare ($\cong$ 1%) and for NMR much less sensitive than either ^{31}P or ^{1}H. Therefore, ^{13}C NMR either is restricted to molecules which are present at very high concentrations or requires an application of ^{13}C-enriched compounds. The spin-spin coupling of ^{13}C and ^{1}H nuclei of the same molecule further complicates the already-complicated spectrum by splitting of the ^{13}C resonance lines. To avoid this effect, special decoupling conditions must be maintained during the measurement.[14] It was proposed quite recently[12,14] to observe ^{13}C nuclei indirectly via ^{1}H NMR, using the above-mentioned spin-spin interactions.

In spite of numerous ^{13}C NMR experiments (recently reviewed by Williams and Gadian[14]) and of the many compounds suggested for intracellular pH measurement, no such measurement has been done. Even though bicarbonate-exchange kinetics was already studied *in vivo* by NMR,[22] it was not attempted to estimate intracellular pH.

D. ^{14}N NMR

The ^{14}N spectra give generally very broad peaks and, therefore, are not very useful. The only exception are molecules of tetrahedral symmetry around the nitrogen atom (e.g., choline, betaine, NH_4^+, and several amino acids).[23] So far, none of these substances has been used for intracellular pH measurement.

E. ^{15}N NMR

The inherent sensitivity of ^{15}N is much lower than that of ^{14}N or ^{13}C, but the clarity of

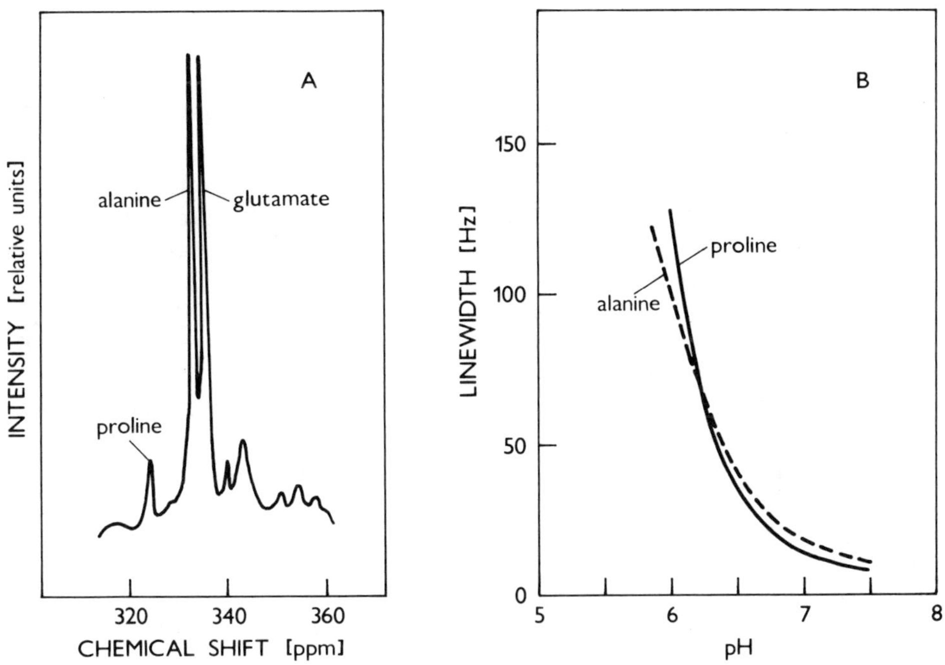

FIGURE 63. (A) A section of a ^{15}N NMR spectrum corresponding to the resonances of proline, alanine, and glutamate inside the cell of *N. crassa* (reference was nitric acid); (B) the *in vitro* pH dependence of the half-width of the resonance peak of proline and alanine (the "linewidth"). (Redrawn from Reference 24.)

biological ^{15}N spectra makes intracellular pH measurement with ^{15}N feasible. Two techniques with amino acids are in use, based on the chemical shift and on the linewidth of the peak, as described below.

1. Histidine

The N3 (π)- and N1 (τ)-nitrogens of histidine (p$K \cong 6$) exhibit very substantial chemical shift changes in the pH range between 4 and 8 and, thus, were suggested for intracellular pH measurement of slightly acidic environments.[24] The feasibility of such measurements was demonstrated in *Neurospora crassa* (Figure 63).

2. Alanine and Proline

Also, the linewidth of proton-coupled ^{15}N resonances of α-amino groups of amino acids and the guanidino group of arginine can be employed as a pH-sensitive NMR parameter. Legerton et al.[24] found that the resonance peak of alanine and proline gradually sharpens over the pH range from 6 to 8 (Figure 63B) and used this phenomenon for the selective determination of cytoplasmic (proline, alanine) and vacuolar (histidine) pH in *Neurospora*. Their interpretation is based on the assumption that the line-broadening of the proton-coupled ^{15}N resonances of intracellular arginine, alanine, and proline is caused predominantly by proton exchange.

In all the above measurements, the reference compound was external HNO_3.

F. ^{19}F NMR

^{19}F is an attractive nuclide for NMR studies because of its high sensitivity, approaching that of ^{1}H, and almost 100% natural abundance. However, as fluorine is rarely present in biological molecules, there is a necessity to use extrinsic fluorinated compounds for intracellular pH measurement.

1. Fluorinated Amino Acids

The technique was introduced by Taylor et al. in 1981.[25,26] They used mono-, di-, and trifluorinated amino acids which penetrate the cell membrane relatively easily when esterified. Inside the cell, the methyl ester is cleaved to the free amino acid, which then remains trapped inside the cell.

Some compounds did not need the reference of trifluoroacetate and offered the advantage of containing their own internal standard. This was the case of α-difluoromethylalanine, the spectrum of which consists of four peaks (due to fluorine-fluorine coupling), and the difference between the central two of the four peaks is pH-sensitive over the physiological range, yielding a shift of about 0.09 ppm in response to a change of 0.1 pH units.[27,28]

First, experiments were done on erythrocytes,[25] lymphocytes,[26,29] and hepatocytes;[30] then, on frog skin[27] and, recently, on rabbit colon and frog skin tissue.[28] Simultaneously, a number of thorough tests of the technique were performed, including toxicity;[28] and the effect of monovalent and divalent ions[27] and the reliability of ^{19}F intracellular pH estimates were confirmed by comparison with DMO distribution, pH microelectrode measurement, and ^{31}P NMR phosphate signal.[28,29]

G. ^{31}P NMR

1. Phosphorus

Phosphorus (with almost 100% natural abdundance of ^{31}P) is a constituent of numerous compounds present at high concentration inside the cell. Fortunately, the very compounds that are abundant and important for the monitoring of physiological processes are also suitable for intracellular pH estimation. Thus, the ^{31}P NMR measurement of pH is always connected with monitoring of the physiological state of cells.

The number of papers dealing with ^{31}P NMR measurement of intracellular pH is really impressive, and the technique has become a routine one.[31-90] One example for all, a special NMR life-support incubator unit is clinically used for monitoring neonatal brain functions, where an intracellular pH shift is employed as an alarming signal of the onsetting brain damage.[31]

The phosphate group is the most popular with intracellular pH estimates of inorganic phosphate, ATP, ADP, and phosporylated sugars (Figures 64 and 65).

Occasionally, other compounds such as 2,3-diphosphoglycerate[51] are employed.

The novel applications of the extrinsic membrane-permeant pH-sensitive compound methyl phosphonate allow not only the simultaneous estimation of both intracellular and extracellular pH (the extracellular signal can be quenched by addition of $MgCl_2$) based on the values of chemical shift,[32,33,54] but also calculation of the Donnan ratio and, thus, of membrane potential.[32,33]

2. Orthophosphoric Acid

Orthophosphoric acid (often simply referred to as inorganic phosphate or P_i) undergoes dissociation according to

$$H_3PO_4 \leftrightarrows H_2PO_4^- \leftrightarrows HPO_4^{2-} \leftrightarrows PO_4^{3-}$$

All four forms have unique electronic structures and, thus, produce different chemical shifts. The titration curve closely follows the shape of a theoretical Henderson-Hasselbalch curve.[36,56,71,72,85]

The apparent p*K* value of the middle transition is 6.75 to 6.9 and, therefore, sensitive preferentially in the pH range of 6 to 8. The separation between the two peaks is about 2.5 ppm ($H_2PO_4^-$ is 0.6 ppm relative to H_3PO_4 and 3.3 ppm relative to phosphocreatine, while HPO_4^{2-} is 3.1 ppm relative to H_3PO_4 and 5.8 ppm relative to phosphocreatine).[51,52]

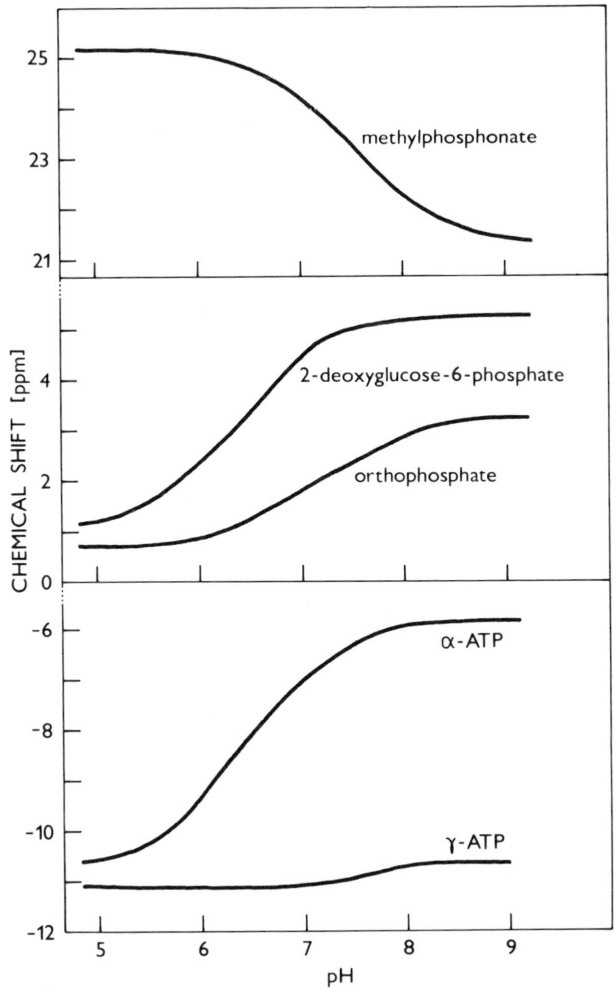

FIGURE 64. A schematic comparison of the pH dependence of chemical shifts of most commonly used phosphorus compounds (at low ionic strengths of the medium). (Compiled from References 13, 52, and 67.)

The ^{31}P NMR signal line-width inside the cell is quite broad (30 to 50 Hz at 145 MHz),[52] so that intracellular pH differences smaller than 0.2 pH units may be difficult to follow.

The dependence of the observed chemical shift on ionic strength and on cation and protein concentration is discussed in Chapter 4E. III. G.

3. Adenosine Triphosphate (ATP) and Adenosine Diphosphate (ADP)

The transition $ATP^{2-} \leftrightharpoons ATP^{3-}$ has a pK of about 6.6 and can also be employed for intracellular pH measurement. The ^{31}P NMR spectrum consists of three separated peaks corresponding to each phosphate group. The peaks of ADP practically coincide with those of α and β phosphates of ATP[51,70,84] (cf. Figure 77). The chemical shifts of each phosphate group show a different pH sensitivity (α-phosphate varies between 5.5 and 10.5 ppm; the β-phosphate is much less pH sensitive [Figure 64] and the γ-phosphate is not used for pH measurement at all). Actually, none of the ATP peaks is a reliable pH indicator as the position of the peak and its shape (splitting) depends also on the conformation of the molecule, which presumably varies inside the cell.[13,53]

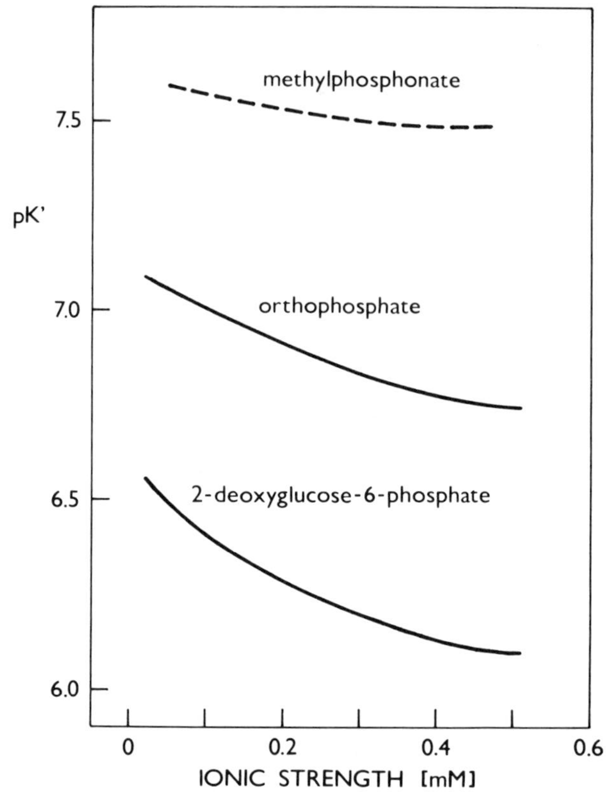

FIGURE 65. A schematic comparison of ionic-strength-induced p*K* shifts of selected pH-sensitive phosphorus compounds. (Compiled from References 32, 52, and 67.)

4. Methyl Phosphonate

It is an extrinsic membrane-permeant compound with pH-indicator properties based on the transition $CH_3PO_3^{2-} \leftrightharpoons CH_3HPO_3^-$, with a p*K* of 7.65 at physiological ionic strength. The corresponding change in chemical shift (relative to H_3PO_4) from 25 ppm of the monobasic form to 21 ppm of the dibasic form is approximately 1.7 times greater than the difference in chemical shifts of orthophosphate. The pH standard curve is reportedly independent of ionic strength.[52]

The membrane permeability of both charged forms of methylphosphonate, demonstrated in erythrocytes[32-34,54] and bacteria,[55] allow the membrane potential as well as the intracellular and extracellular pH values[32,54] to be estimated.

5. Phosphorylated Sugars

The accumulation of sugars inside the cell and their subsequent phosporylation may eventually lead to high intracellular concentrations of phosphorylated sugars. Their pH-dependent ^{31}P NMR shifts can easily be used for intracellular pH estimation. In this way, intracellular pH was measured with glucose-6-phosphate,[50,76,77] fructose-1-phosphate,[37] fructose-2-phosphate,[50] or 2-deoxy-D-glucose-6-phosphate[35,57,58,73,86] after incubation of cells in glucose, fructose, or 2-deoxyglucose, respectively.

6. Reference Compounds

The reference compound is usually external H_3PO_4 or internal phosphocreatine. Nevertheless, the use of inorganic phosphate as an internal reference for, e.g., methylphosphonate,

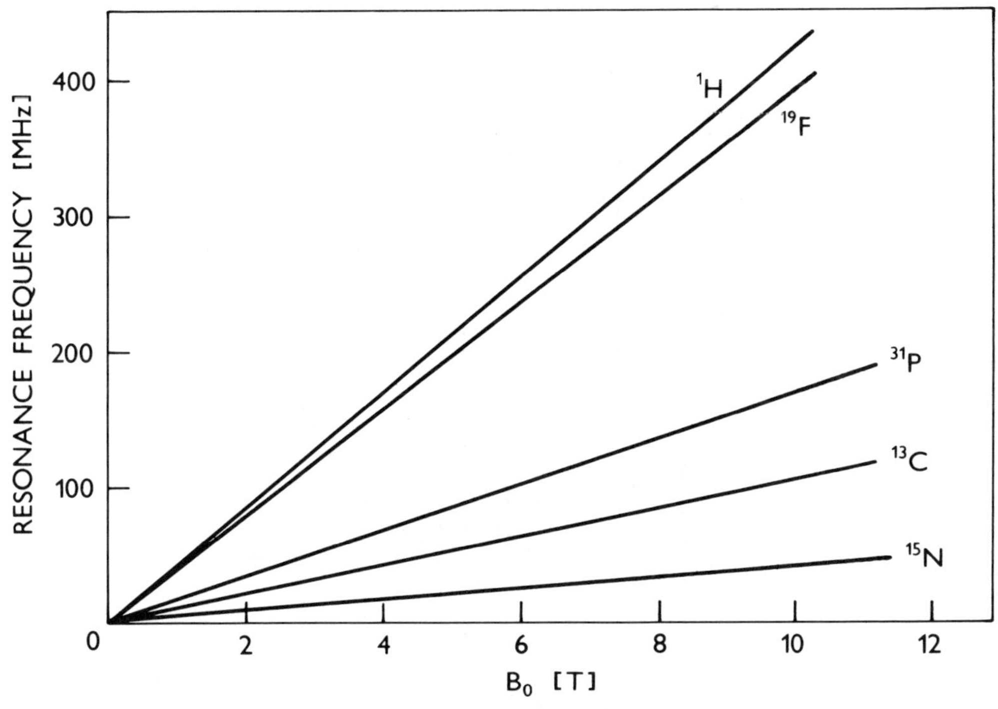

FIGURE 66. Dependence of the nuclear resonance frequency on the intensity of magnetic field for different isotopes used in intracellular pH measurement.

is also worth considering as this combination may compensate the effects of medium composition of pK and chemical shift value.[32,55]

III. EXPERIMENTAL ASPECTS

A. NMR Spectrometer

An NMR device employs two magnetic fields oriented perpendicular to each other, viz., the strong and homogeneous steady magnetic field B_0 and the weaker observing magnetic field B_1 oscillating with a high frequency (in the radio-frequency range).

The steady magnetic field B_0 is usually produced by a permanent magnet or electromagnet (typical intensities achieved are 0.5 to 3 T) or, better, by a superconducting magnet (field range 4.8 to 14.4 T). According to Equation 154 and Table 6, the corresponding nuclear resonance frequencies lie in the MHz range (Figure 66). Therefore, the observing field is effected by a radio-frequency coil, and the NMR signal is detected by a similar coil (part of a resonance circuit) in the direction perpendicular to both B_0 and B_1 (Figure 67).

The standard NMR apparatus, pulsed FT NMR spectrometer, costs around \$200,000 to \$600,000 (U.S.) (e.g., made by Varian, Bruker, Jeol).

B. Strategies for Localization of the Signal

1. Introduction

In biological experiments, including pH measurement, it is often important to differentiate the signal from different parts of the sample (body, organ, tissue). Using topical magnetic resonance and surface coil technique, the average pH values of chosen organs or tissues can be derived, while the NMR imaging already offers the possibility of the derivation of pH topography of the tissue, though with relative lower spatial resolution than optical techniques.

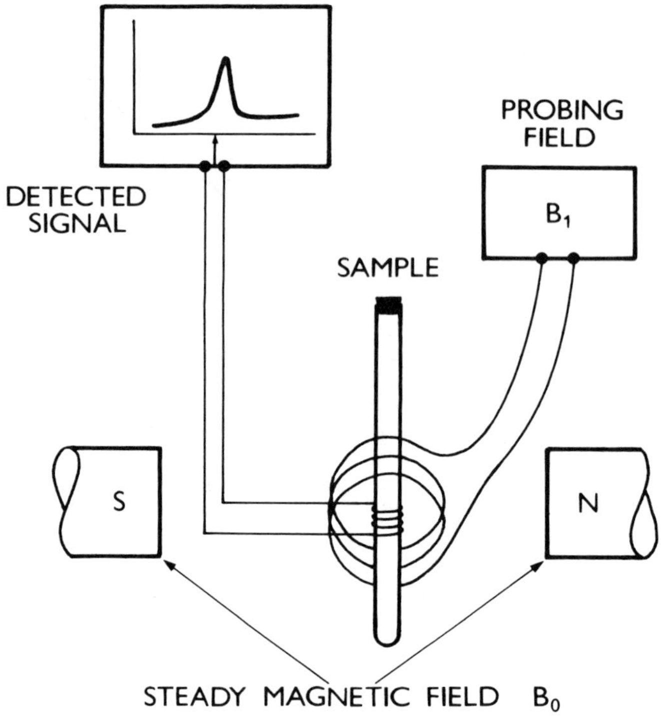

FIGURE 67. A basic scheme of a NMR spectrometer.

In the usual arrangement, the whole sample is exposed to both magnetic fields, B_0 and B_1, and the signal is collected over the whole volume of the sample. The modifications mentioned above restrict the area of interest from which the NMR data are collected, either by suppressing or degrading the signal from the rest of the sample (surface coil technique and the topical magnetic resonance) or by differentiating the signal according to the place of its origin (NMR imaging).[12,14]

2. Topical Magnetic Resonance (TMR)

This technique[12,59,60] uses a sophisticated magnet system which produces a homogeneous magnetic field only inside a specified region. The whole sample is exposed to a radio-frequency field B_1, and the NMR signal is collected over the whole volume. However, a sharp signal can be obtained from the area of homogenous magnetic field only; the remaining areas outside the homogenous region yield very broad lines resulting in an enhanced background. Though the background signal can be suppressed by special techniques, the poorer signal-to-noise ratio may cause problems.

3. Surface Coil

This technique[61,82] places the whole object in the steady homogeneous magnetic field B_0, but only a small part of the object is exposed to B_1. This is done by means of a special coil in the form of a simple planar loop placed over the volume of interest. The signal is acquired from a disc-shaped tissue of about one coil radius thick. In comparison with the topical magnetic resonance, this technique has a higher signal-to-noise ratio. It was quite successful in investigating superficial tissues but did not work so well with deeper-located tissues and organs. In such cases, small coils were put into a body cavity (e.g., stomach, bladder) or simply inserted inside the tissue of interest.

4. NMR Imaging

Both the surface coil technique and the topical magnetic resonance have now been largely superseded by techniques based on imaging principles, also called NMR tomography or zeugmatography.[5]

The NMR imaging apparatus uses the pulsed NMR spectrometer, Fourier transformation, and a sophisticated computer. The whole sample is put inside the coil. The spatial localization is achieved by a superimposed, strong, linear magnetic field gradient onto homogeneous steady magnetic field B_0. The supplementary field is slowly varied so that one element of the sample ("pixel") after another comes into the resonance conditions. The fully three-dimensional images are derived by computer from pictures of separated planes (slices) or from several adjacent planes at a time (multislices). Because of this apporach, which saves both the capacity of the memory and the processing time, such devices are usually called tomographs (from the Greek *tomos* — a cut).

At the present technical state-of-the-art of NMR imaging, the signal from each pixel is evaluated according to its amplitude and/or T_1 and/or T_2. The apparatus that would give a complete NMR spectrum from each pixel has not been constructed yet,[12] though in theory it could be easily accomplished; but in practice technical difficulties are considerable. The main problem seems to be the broadening of spectral lines due to the application of the magnetic field gradient. To solve this, special pulse experimental techniques had to be developed.[5] So far ^1H imaging is routinely used in medicine (differentiation based on T_2), and several ^{31}P imaging devices are under construction. The typical concentration threshold of these ^1H imaging techniques is around 10 m*M*.[12] The quality of the picture is usually low, approaching that of X-ray photography and not allowing distinction between individual cells in the tissue.

C. Techniques of Measurement

In principle, the amplitude of the observing field B_1 may be either constant (continual wave techniques, cw) or gated (pulse techniques). In the former arrangement, the magnetic resonance signal is detected continuously; in the latter case, only in the time between pulses.

The continual wave techniques are rarely used in biology, where they were superseded by significantly more sensitive pulse techniques.

In different versions of pulse techniques, various combinations of radio-frequency pulses are employed. The NMR signal represents the response of nuclear spins after termination of the pulse or pulse sequence, its nature depending on the duration of pulses and the time intervals between them.[90]

The simplest pulse NMR experiment is based on an application of a single pulse of duration τ fulfilling the equation

$$\gamma B_1 \tau = \pi/2$$

the so-called $\pi/2$ or 90° pulse.

The response (transverse magnetization signal) is called a free-induction decay (FID) curve and resembles finely modulated damped oscillations with amplitude gradually diminishing according to e^{-t/T_2} (Figure 68). The FID curve is easily transformed into the familiar shape of an NMR spectrum by a mathematical procedure called Fourier transformation. From the mathematical point of view the two curves, FID and NMR spectrum, are equivalent. They represent only two different forms of the same thing, in the "time domain" (FID), the other case in the "frequency domain" (spectrum). The Fourier transformation merely converts one image into another.

The acquisition of a single FID following just one exciting pulse would, in principle, be sufficient for a complete NMR spectrum. However, as noise obscures the fine shapc of the

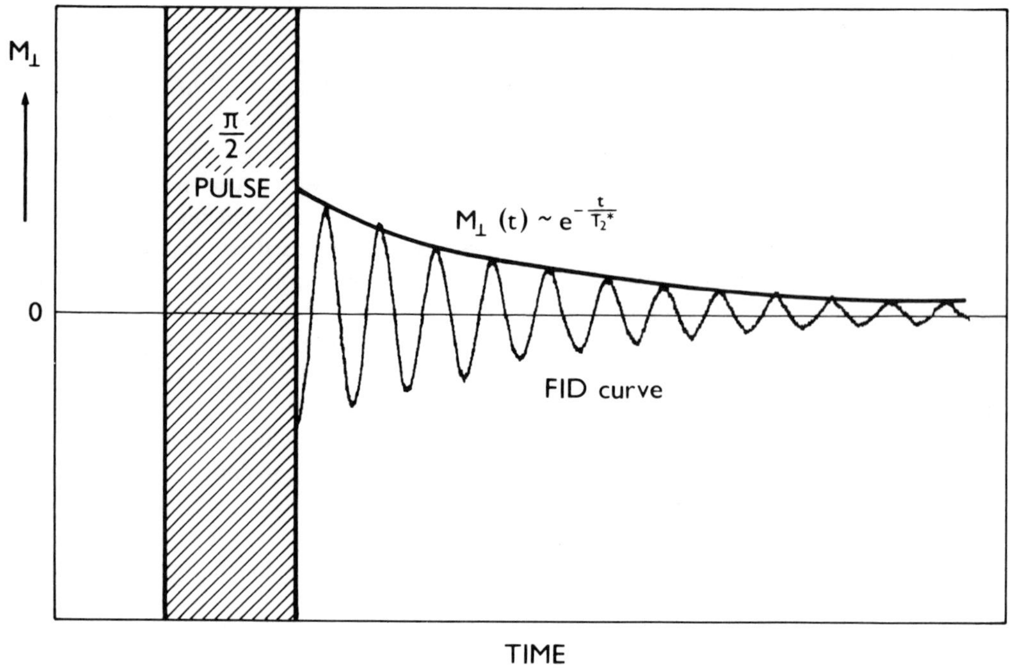

FIGURE 68. A typical free induction decay (FID) curve. ($M_{\perp}$, corresponding to the value of magnetic induction, forms an envelope of the decay curve as expressed by the formula shown in the figure.)

FID curve which carries the information needed for the determination of the NMR spectrum, it is necessary to repeat the acquisition procedure (several hundred or thousand times, depending on the signal-to-noise ratio and the spectral resolution required; cf. Figure 69).

The two-pulse technique allows T_1, T_2 to be measured and the recorded NMR spectra with respect to T_1 and T_2 to be modified. In this way, sharp but weak peaks with a large T_2 can be enhanced from the broad background. This procedure is often used in ^{1}H NMR of biological samples, where the weak signal of a pH-indicating substance is buried in the broad signal of 100 *M* background of protons of the biological tissue.

D. Sensitivity and Time Resolution

The sensitivity and the time resolution are the Scylla and Charybdis of NMR of biological samples. An increase in sensitivity (signal-to-noise ratio) is usually possible only if the time of the accumulation of data is increased and vice versa. To make things worse, the measurement of a biological sample requires a substantially higher sensitivity because peaks from inside the cell are usually broader (and, thus, smaller as the peak area is constant) than those from the external medium.[62] For instance, in ^{31}P NMR, the threshold sensitivity 0.1 to 0.3 m*M* (10 μ*M* in the top machines) is consequently shifted to concentrations above 1 m*M*.[52]

The sensitivity can be increased, e.g., in the following ways:

1. Large sample volumes (tubes up to 20 mm in diameter) are preferred.
2. High densities of cells are used.
3. High magnetic fields from superconducting solenoids increase the sensitivity with approximately the square value of B_0 (more precisely $B_0^{7/4}$).[5]
4. The differences in T_2 can be employed to distinguish weak, sharp peaks on a broad background (e.g. the SEFT NMR technique).

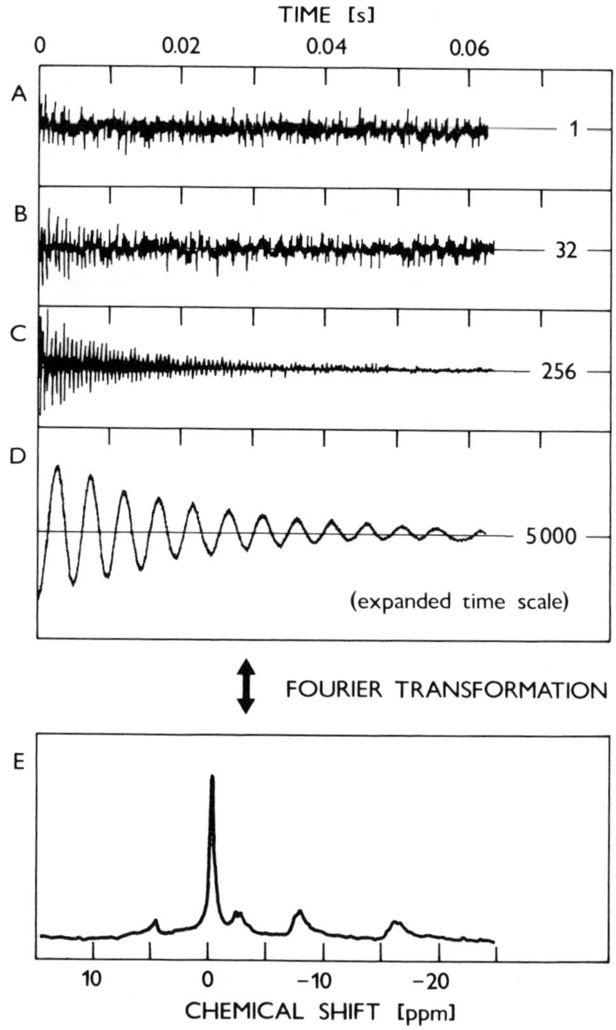

FIGURE 69. Schematic representation of the effect of accumulation of the FID signal (obtained after 1, 32, 256, and 5000 pulses; A, B, C, and D, respectively) and the Fourier transformation (E) of the resulting signal into the shape of an NMR spectrum. (Adapted from Reference 12.)

The usual pH-measuring experiment requires 1 to 30 min to accumulate enough data for an NMR spectrum. A typical value for 5 m*M* ^{31}P is about 1 min for a S/N ratio = 5.[52] However, this is too slow for the study of dynamic aspects of metabolism, when the temporal resolution should be at least one order of magnitude shorter than the duration of the metabolic event observed. With cyclic events, the temporal resolution can be significantly increased when the timing of the probing pulse is synchronized consecutively to specific time periods within the cycle, and if the cycle is repeated several times. In this way the time resolution of seconds,[91] or even down to 2 ms,[92] was achieved. The theoretical time resolution limit is then given by the relaxation times of nuclei, e.g., a few milliseconds for ^{31}P.

E. How to Prepare a Biological Sample

The NMR technique is a truly nondestructive and noninvasive method in the way that you can put your own limb or your whole body inside the magnet of the NMR machine without endangering your health,[93] and the magnetic field is harmless at the intensities used in NMR as confirmed by the U.S. F.D.A.[94,95] Limitations imposed on the sample by the measurement process per se are minimal. The only problems that may arise are those of maintaining good physiological conditions of a high density of cells during the measurement. Due to the low sensitivity of NMR, cell concentrations of up to 20 to 50% are employed. The maintenance of adequate oxygenation, supply of nutrients, and possible removal of toxic metabolic by-products often require considerable effort. However, NMR (especially ^{31}P NMR) offers a unique opportunity to monitor simultaneously both pH and physiological processes and, in this way, to indicate a possible deterioration of the sample.

The basic way to prepare an NMR sample of suspension cells is to cool freshly harvested cells down to about 10°C, then centrifuge and wash twice with ice-cold buffer (preferably, a simple mineral buffer impermeant in the cell). The temperature is then raised to the desired value before the NMR data acquisition starts. This procedure can be applied in principle to all kinds of cell suspension (microorganisms, blood cells, tumor cells).

Cells that require attachment to an appropriate substratum are grown on beads, hollow fibers, or in special cell culture units (e.g., with polymeric membranes inside) designed to fit in the solenoid coils. The purpose of such units is usually twofold: first, to increase the surface on which cells grow and, thus, the cell density; and, second, to permit easier aeration and perfusion of cells. For inspiration, see, for instance, Reference 96. The material on which cells are grown should be chosen with respect to its magnetic properties as the very presence of the sample perturbs the homogeneity of the magnetic field and makes the NMR lines broader. For instance, the use of hydrated beads is discouraged since the intra-bead volume has a significantly different magnetic susceptibility than the bulk medium.[52]

1. Stirring

In contrast to optical techniques, sedimentation of cells does not affect NMR measurements. However, in order to maintain a more even aeration of cells, stirring in combination with bubbling is often used. A conveneint way may be to use a magnetically driven stirring bar (Figure 70).

2. Aeration

Simple bubbling of oxygen through a (possibly stirred) suspension of cells may have two unpleasant consequences.

First, a transient perturbation of the homogeneity of the magnetic field appears. This is avoided by collecting NMR data exclusively in the time between two consecutively released bubbles. The lock signal monitoring the homogeneity of magnetic field can be employed for intermittent gating of the accumulation of NMR signals. When the system returns to equilibrium after the bubble, the accumulation of data is resumed.[62]

Second, bubbles may mechanically disturb some more sensitive cells, or even cause their lysis. Therefore, in such cases very gentle bubbling or perfusion is recommended.

3. Perfusion

The perfusion of a sample is relatively simple to achieve, and many reliable flow-through, perfusion-type systems have been developed and described in the literature[52,63,96-99] (Figure 71). The streaming should be adjusted to the fragility of the cellular species used.

Difficulties may arise with suspensions of smaller cells which tend to flow away with the perfusate. This can be solved by immobilizing the cells, which can be done *in vivo* in various ways (recently reviewed by Mattiasson[100]):

1. Entrapment in inert semipermeable materials such as hydrogels, fibers, or membranes
2. Adsorption on an inert, solid support
3. Covalent binding, including cross-linking to an activated support

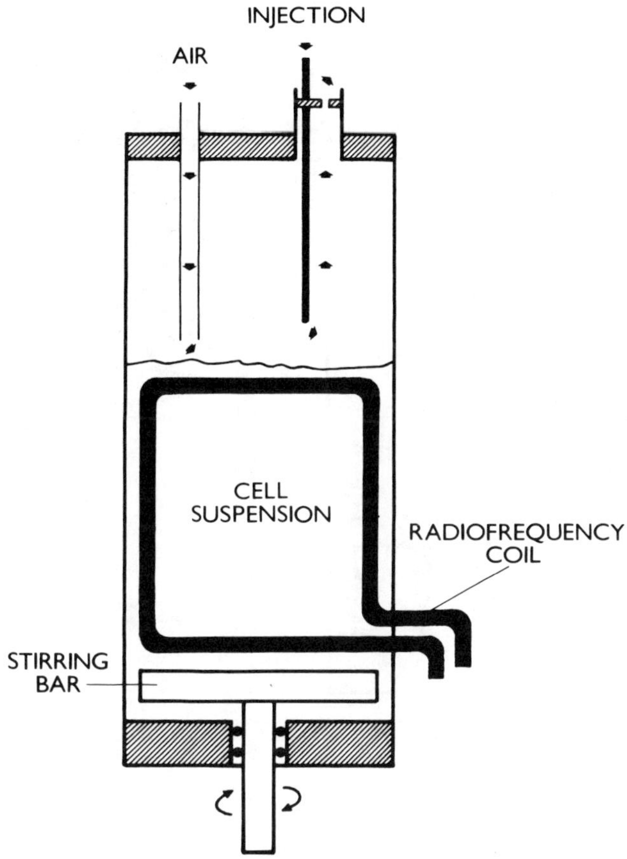

FIGURE 70. Stirring and aeration of a cell suspension in a NMR arrangement.

A convenient example of the first group is mixing cells with warm agarose, dropping to low temperatures, and then extruding the mixture through a cooled capillary tube; or forming small capsules with immobilized cells in a mixture of alginate and calcium chloride.[97-100]

A solution to the second problem is to attach cells to beads (e.g., polylysine-coated Sephadex® beads). However, cells may gradually separate from the beads during perfusion and flow away.

F. Interpretation of Results

The interpretation of NMR spectra may be even more complicated than their acquisition. The essential step is a correct structure assignment of resonances. It should be done very carefully and cautiously by an experimenter with great experience in both theory and experiment. ^{31}P NMR spectra are simpler and, thus, easier to interpret. The more complex ^{13}C and ^{1}H NMR spectra represent a greater challenge in making the assignment. In ^{31}P NMR, it may be expected that compounds present at concentrations lower than approximately 0.2 to 1 m*M* cannot produce an observable peak. On rare occasions one peak may be masked by that of another compound with a similar chemical shift (e.g., ADP and NAD in ^{31}P NMR).

Unfortunately, all peaks of intracellular substances are wider than those in solution by approximately one order of magnitude[11, 52] (Figure 61). The origin of this intracellular line-broadening is unclear[11] (see chapter on subcellular-pH heterogeneity for details). The higher

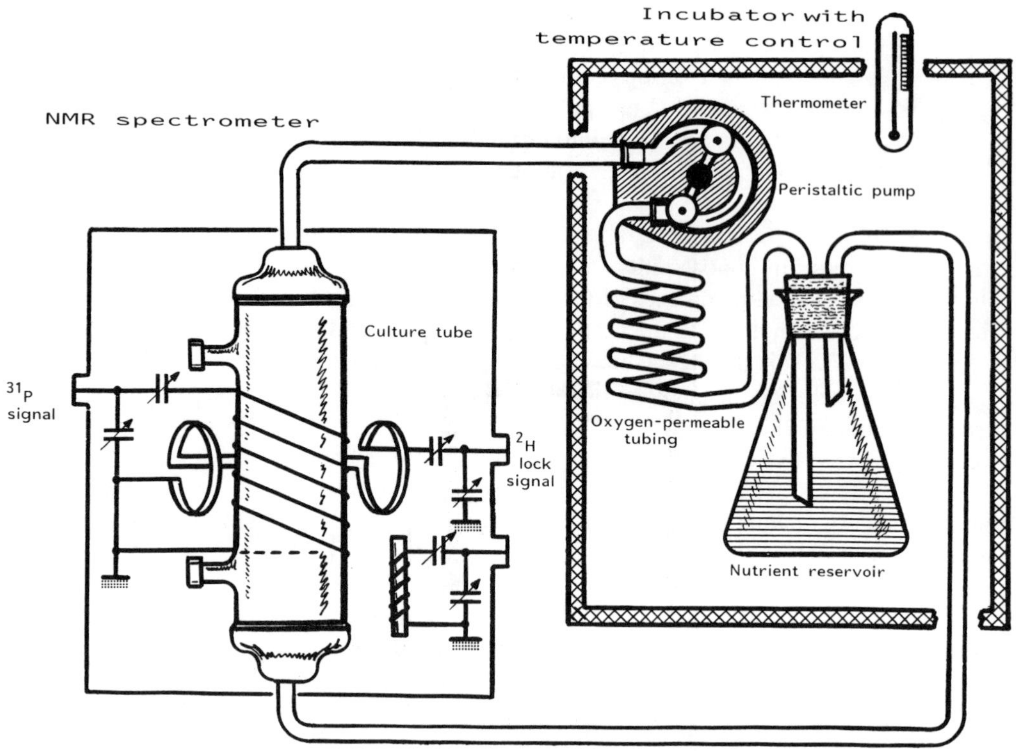

FIGURE 71. Schematic drawing of a continual perfusion system in NMR. (Redrawn from Reference 100.)

width of the peaks results in their lower height. Therefore, a small peak can be easily buried in the broad background. This contributes not only to lower sensitivity, but also to a more complicated assignment of peaks.

G. Problems of pH Calibration

1. Introduction

Like any other technique, there are, unfortunately, many factors besides pH that may influence the measured NMR data and result in an erroneous estimate of the intracellular pH value.

Calibration should be performed as described in general in Chapter 4A. However, the use of NMR for monitoring of pH-indicating substances brings specific calibration problems, which, with respect to the fundamental Equation 163, can be divided into those affecting the value of apparent p*K* and those distorting the value of c_B/c_A.

Any shift in apparent p*K* will result in the same shift in the measured pH value (Equation 163). The problems of stability of p*K* are common to all pH-indicating substances and are discussed in detail in Chapter 4A.

The measured value of the chemical shift is influenced by contributions of other substances than the dissociated form of a pH indicator and its conjugate base. In practice, many components of intracellular medium do form complexes with the pH indicator. Such complexes are often transient (they interchange with each other very rapidly) and have chemical shift values slightly different from those of the free form of the indicator (cf. Chapter 4E. I. G. 1 exchange reactions).

For instance in ^{31}P NMR, the divalent metal ions and basic proteins and amino acids form complexes with pH indicators of the phosphate type. Furthermore, large pH-indicating molecules, such as ATP, exhibit different chemical shifts depending on the attained conformation of the molecule.[13,53]

Therefore, instead of the two postulated interchanging forms of the indicator, in practice, the peak of the pH indicator results from a large number of merged resonance peaks, arising from forms A and B and their various complexes and conformations. So, instead of Equations 163 and 164, Equation 165 must be applied. Unfortunately, both the concentrations c_i and the chemical shift values δ_i are difficult to estimate, which further complicates the situation.

To make things worse, the mean concentration of complexes varies with the concentration of competing substances. The above-mentioned formation of complexes of metal ions with phosphates is characterized by competition of all phosphate compounds for metal ions. Hence, an increase in concentration of any phosphate compound results in a decrease of concentrations of metal complexes of the pH-indicating compound. Consequently, the contribution of these complexes to the NMR signal decreases, and a different value of chemical shift is measured. This difference can then be erroneously interpreted as a pH change.

In practice, the effect of the composition of a cytoplasmic medium causes an uncertainty in pH estimation of about 0.2 to 0.5 pH units, in extreme cases up to 2 pH units. The effects of composition on the value of the chemical shift measured are so complex that it is often preferable to do the pH calibration directly inside the cell under *in vivo* conditions or in cell homogenates. The nature of these effects was studied mainly for ^{31}P NMR, so the following data are typical of inorganic phosphates, sugar phosphates, ATP, ADP, and similar compounds.

2. Effect of Ionic Strength

The effect of ionic strength on the apparent p*K* is clearly established as discussed in the chapter on pH indicators. According to the theory of electrostatic interaction, it should apply in a similar way to all the pH-indicating substances used. However, gray is the theory, and there are clearly individual differences between compounds. The effect simply ascribed to ionic strength appears to be a combination of two different effects. First, it is the purely electrostatically induced p*K* shift, and second, some kind of additional and specific effect of ionic strength on the measured value of the chemical shift.

This is clearly illustrated on the case of inorganic phosphate and glucoso-6-phosphate in the difference between Figures 72 and 73. The first picture shows the p*K* shift due to the addition of 0.1 and 0.5 *M* KCl (as observed by potentiometric titration), whereas the second picture shows the same effect as it appears in the NMR chemical shift. By comparison of these curves, it is evident that the apparent p*K* shift observed via chemical shift comprises some additional effects besides the pure electrostatic p*K* shift. These additional effects may even compensate the electrostatic effect and make p*K* seemingly independent of ionic strength (compare, e.g., methylphosphonate with phosphate compounds).[52]

3. Effect of Divalent Cations

Virtually all phosphate compounds employed in ^{31}P NMR form complexes with metal ions resulting in chemical-shift changes, the most pronounced effect being observed in the case of esterified phosphates, such as ADP and ATP. In these two compounds, false pH changes of up to 2 pH units may appear. Among the divalent metal ions, it is especially magnesium and calcium which are potentially important because of their essential presence in all living cells. The concentration of free Mg^{2+} is higher (~m*M*) than that of free Ca^{2+} (less than ~1 μ*M* in cytoplasm or ~100 μ*M* in mitochondria).

The effect of divalent metal ions is twofold. First, their presence increases considerably the ionic strength and, thus, shifts the apparent p*K* (Figure 74), as described previously. Second, the contribution of the signal arising from metal complexes falsely alters the measured value of the chemical shift (Figure 75).

Let us consider a very simple model based on phosphate and Mg^{2+} in the solution.[67] The NMR signal arises from four phosphate species rapidly exchanging with each other

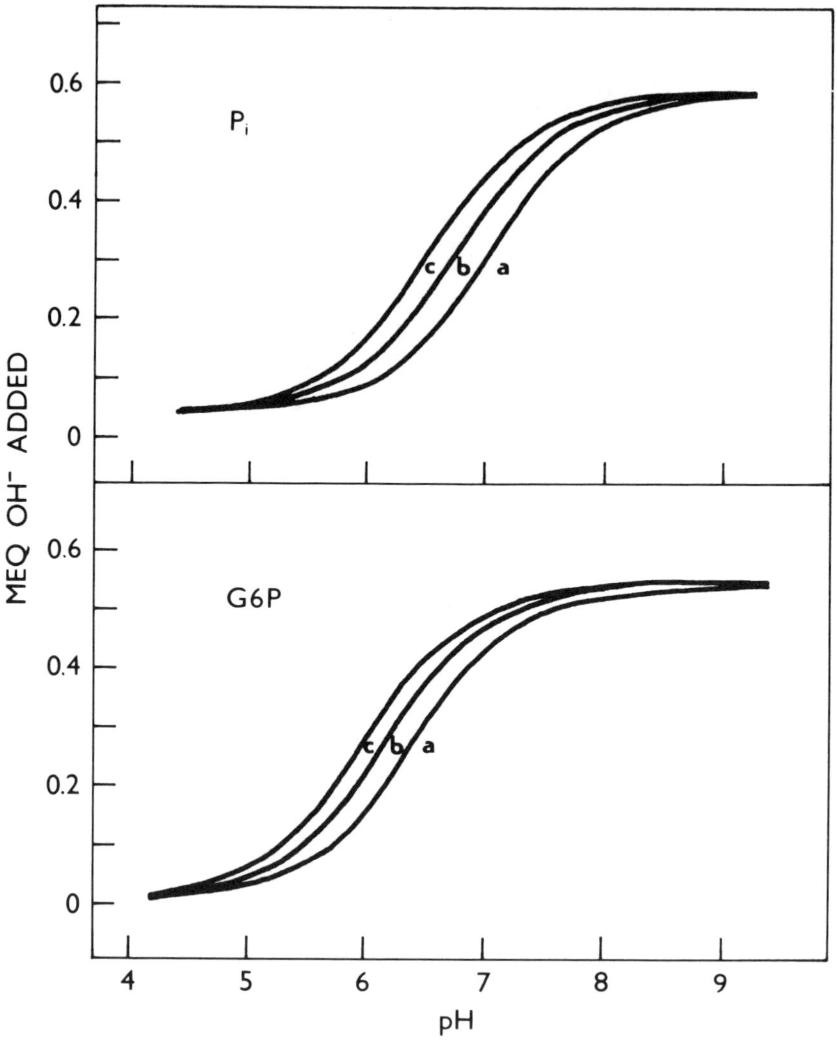

FIGURE 72. Potentiometric titration curves of 5 m*M* potassium phosphate (above) and 5 m*M* glucose-6-phosphate (below). Curves a: the compounds alone; curves b: with 0.1 *M* KCl; curves c: with 0.5 *M* KCl. (Adapted from Reference 67.)

(HPO_4^{2-}, $H_2PO_4^-$, $MgHPO_4$, $MgH_2PO_4^+$), each of them having a definite value in the chemical shift.[67] In this solution, four equilibria (see scheme in Figure 76) must be considered:

$$K_1' = c_{H^+}c_{HPO_4^{2-}}/c_{H_2PO_4^-} \quad (166)$$

$$K_2' = c_{Mg^{2+}}c_{H_2PO_4^-}/c_{MgH_2PO_4^+} \quad (167)$$

$$K_3' = c_{H^+}c_{MgHPO_4}/c_{MgH_2PO_4^+} \quad (168)$$

$$K_4' = c_{Mg^{2+}}c_{HPO_4^{2-}}/c_{MgHPO_4} \quad (169)$$

(Furthermore, at a pH far from neutrality, equilibria involving H_3PO_4 and PO_4^{3-} should also be included.)

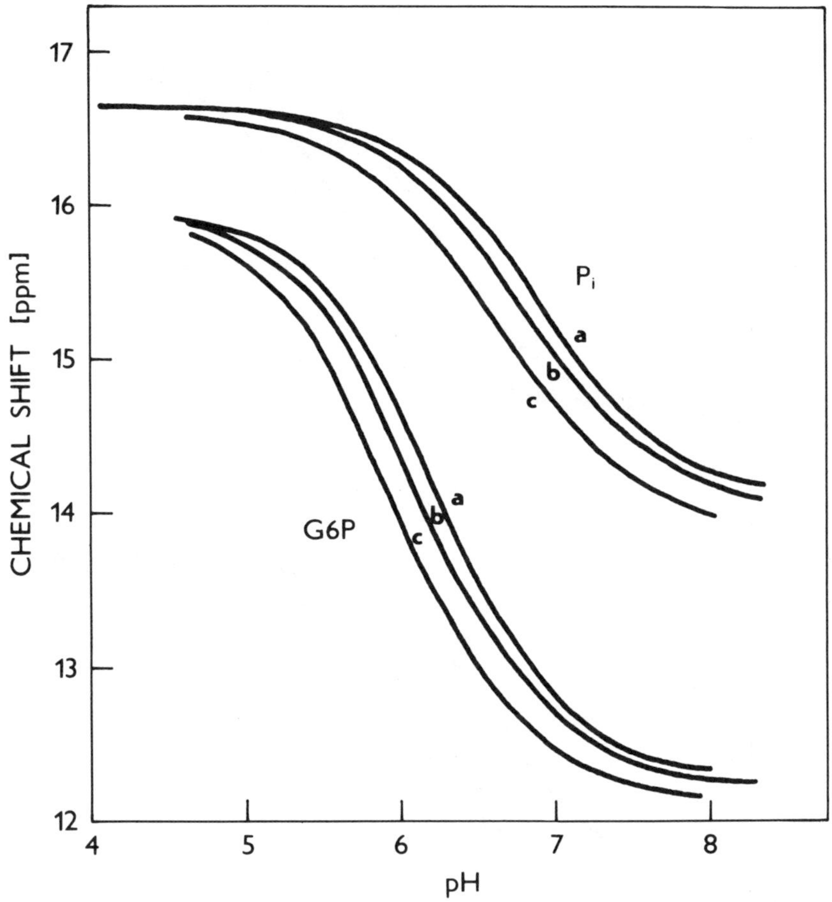

FIGURE 73. Titration curves of 5 mM potassium phosphate (P_i) and 5 mM glucose-6-phosphate (G6P) as in Figure 72 but monitored via the chemical shift. (Adapted from Reference 67.)

According to Equation 164, the observed chemical shift is then

$$\delta = \frac{c_{HPO_4^{2-}}\delta_{HPO_4^{2-}} + c_{H_2PO_4^-}\delta_{H_2PO_4^{2-}} + c_{MgHPO_4}\delta_{MgHPO_4} + c_{MgH_2PO_4^+}\delta_{MgH_2PO_4^+}}{c_{HPO_4^{2-}} + c_{H_2PO_4} + c_{MgHPO_4} + c_{MgH_2PO_4^+}} \qquad (170)$$

or, using above-defined apparent pK constants,

$$\delta = \frac{K_1'K_3'K_4'\delta_{HPO_4^{2-}} + K_3'K_4'c_{H^+}\delta_{H_2PO_4} + K_1'K_3'c_{Mg^{2+}}\delta_{MgHPO_4} + K_1'c_{H^+}c_{Mg^+}\delta_{MgH_2PO_4^+}}{K_1'c_{H^+}c_{Mg^{2+}} + K_1'K_3'K_4' + K_3'K_4'c_{H^+}} \qquad (171)$$

which gives an expression for pH

$$\mathrm{pH} = \log \frac{K_1'c_{Mg^{2+}}(\delta - \delta_{MgH_2PO_4^+}) + K_3'K_4'(\delta - \delta_{H_2PO_4^-})}{K_1'K_3'K_4'(\delta_{HPO_4^{2-}} - \delta) + K_1'K_3'c_{Mg^{2+}}\delta_{MgHPO_4}} \qquad (172)$$

The real situation (Figure 75) is, however, even more complicated than this, as the binding constant is dependent on the ionic strength[13] and, particularly, on the concentration of other phosphate compounds.[67] The competition of other phosphate groups is significant; for in-

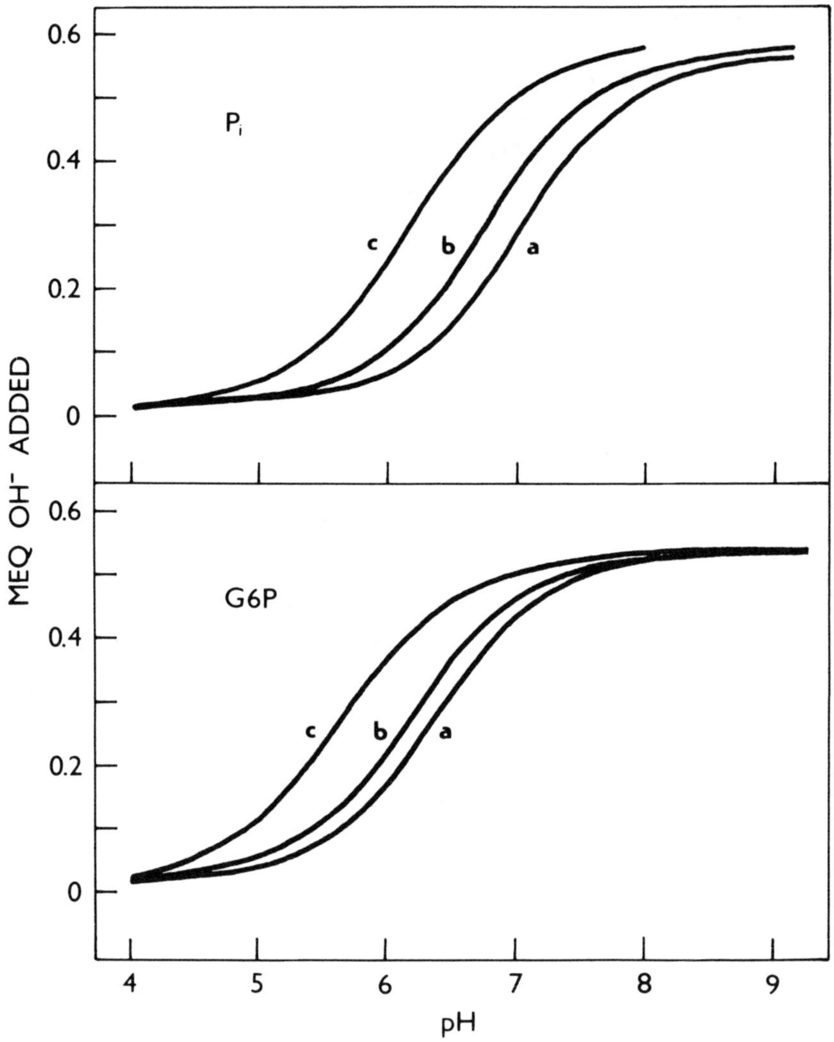

FIGURE 74. Potentiometric titration curves of 5 m*M* potassium phosphate (above) and 5 m*M* glucose-6-phosphate (below). Curves a: the compounds alone; curves b: with 5 m*M* KCl; curves c: with 50 m*M* $MgCl_2$. (Adapted from Reference 67.)

stance, the addition of 5 m*M* ATP eliminated the Mg effect.[67] Besides, estimates of δ_{MgHPO_4} and $\delta_{MgH_2PO_4}$ are rather complicated.

4. Effect of Basic Proteins

The effect of the basic protein protamine on chemical shifts of phosphate and glucose-6-phosphate also looks somewhat mysterious, the influence being apparently rather complex (Figure 77). On the other hand, acidic proteins, such as bovine serum albumin, show little effect.[67]

5. Effect of Basic Amino Acids

The association of phosphates with arginine causes a chemical shift of 0.5 ppm.[101]

H. A Typical Experimental Result

A typical NMR spectrum is illustrated in Figure 78. Any of the substances marked with

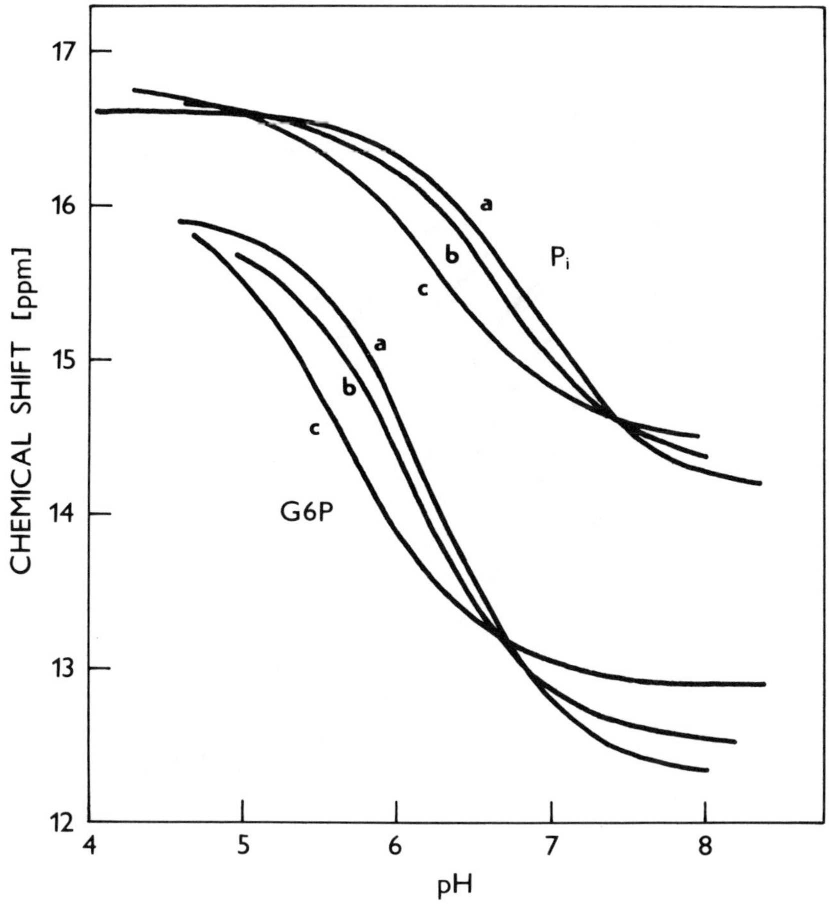

FIGURE 75. Titration curves of 5 m*M* potassium phosphate (P_i) and 5 m*M* glucose-6-phosphate (G6P) as in Figure 74 but monitored via the chemical shift. (Adapted from Reference 67.)

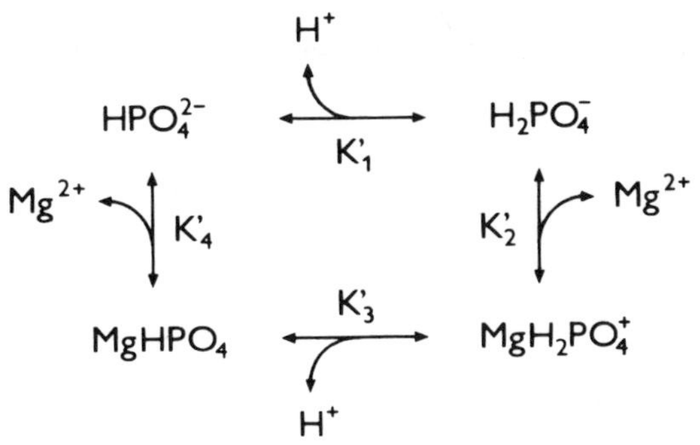

FIGURE 76. The dissociation scheme of orthophosphate in the presence of magnesium ions.

an arrow can be employed for the determination of intracellular pH. However, each of them may yield a different pH_{in} value, because the value measured is an average value of all the milieus in which the particular substance occurs (cf. Chapter 6 on subcellular pH heterogeneity).

The pH value is derived by comparison of the measured value of the chemical shift with that of the standard curve. In the case depicted in Figure 78, the derived intracellular pH

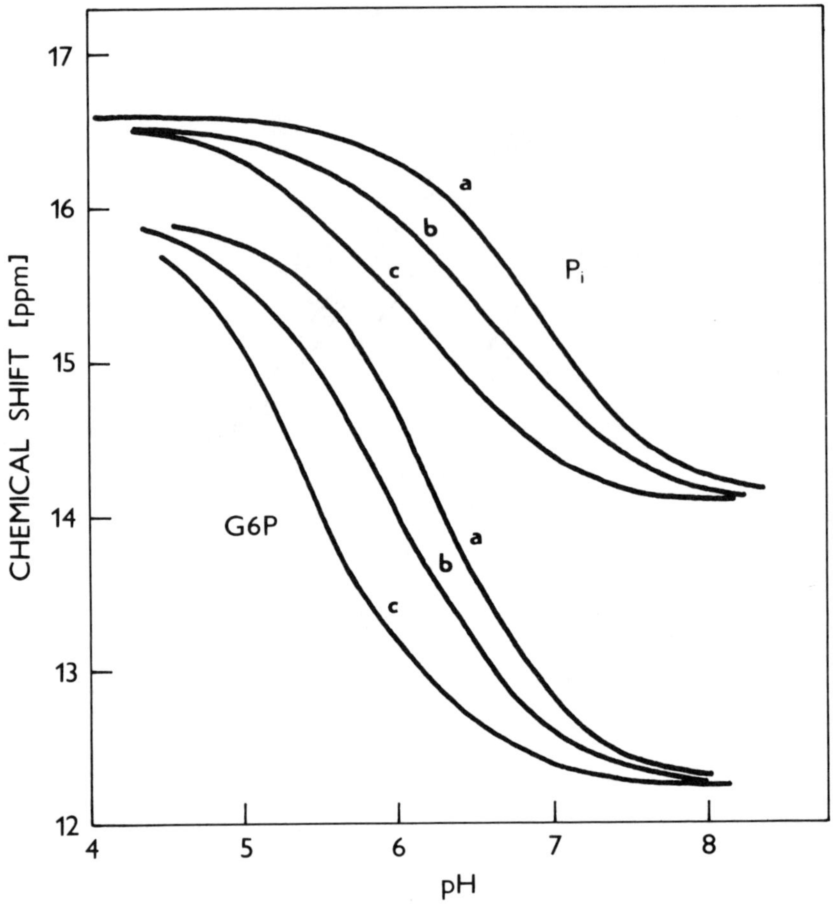

FIGURE 77. Titration curves of 5 m*M* potassium phosphate (P_i) and 5 m*M* glucose-6-phosphate (G6P) alone (curves a); and in the presence of 10 mg protamine per milliliter (curves b) and of 50 mg protamine per milliliter (curves c). (Adapted from Reference 67.)

values are 5.8 and 5.9 for phosphate and ATP with respect to calibration curves on Figures 64 and 73.

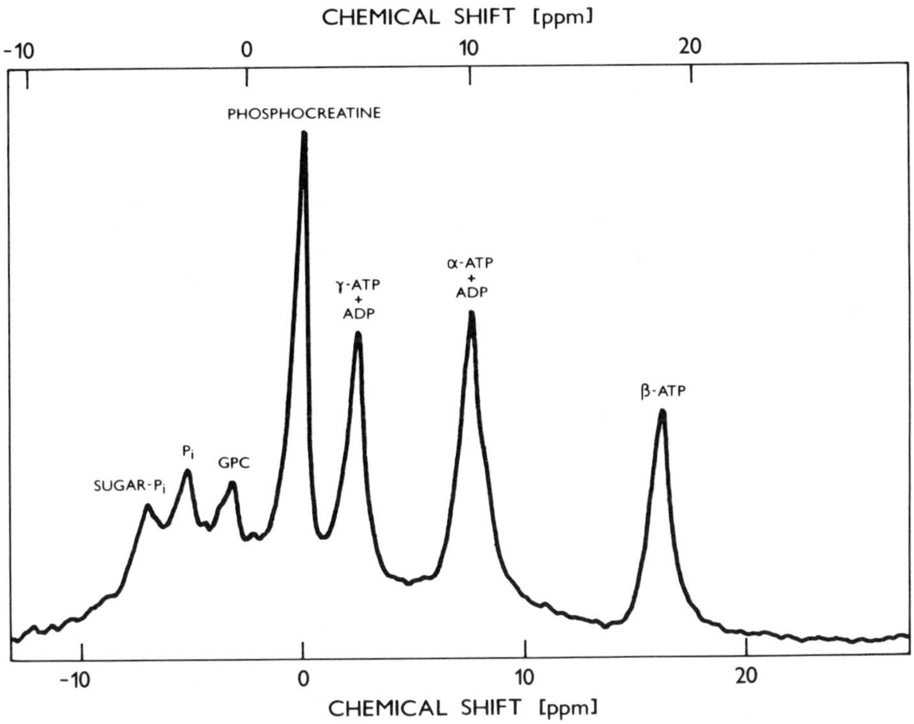

FIGURE 78. A typical example of a simple ^{31}P NMR spectrum of a biological sample (perfused rabbit heart). The observed peaks correspond to the glycolytic sugar phosphates (sugar-P_i), orthophosphate (P_i), glycerol-3-phosphorylcholine (GPC), phosphocreatine, ATP, and ADP (one peak for each phosphate group). The upper scale represents the chemical-shift value expressed in ppm, relative to the phosphoric acid resonance; the lower scale is relative to phosphocreatine resonance.

REFERENCES

1. **Odeblad, E., Bhar, B. N., and Lindström, G.,** Proton magnetic resonance of human blood cells in heavy-water exchange experiments, *Arch. Biochem. Biophys.*, 63, 221, 1956.
2. **Jardetzky, O. and Wertz, J. E.,** Detection of sodium complexes by nuclear spin resonance, *Am. J. Physiol.*, 187, 608, 1956.
3. **Moon, R. B. and Richards, J. H.,** Determination of intracellular pH by ^{31}P magnetic resonance, *J. Biol. Chem.*, 248, 7276, 1973.
4. **Farrar, T. C. and Becker, E. D.,** *Pulse and Fourier Transform NMR; Introduction of Theory and Methods*, Academic Press, New York, 1971.
5. **Gadian, D. G.,** *Nuclear Magnetic Resonance and Its Application to Living Systems*, Clarendon Press, Oxford, 1982.
6. **Mullen, K. and Pregosin, P. S.,** *Fourier Transform NMR Techniques*, Academic Press, London, 1976.
7. **Gupta, R. K.,** *NMR Spectroscopy of Cells and Organisms*, Vols. I and II, CRC Press, Boca Raton, FL 1988.
8. **Becker, E. D.,** *High Resolution NMR: Theory and Chemical Application*, Academic Press, New York, 1980.
9. **Jardetzky, O. and Roberts, G. C. K.,** *NMR in Molecular Biology*, Academic Press, New York, 1981.
10. **Fukushima, E. and Roeder, S. B. W.,** *Experimental Pulse NMR: A Nuts and Bolts Approach*, Addison-Wesley, New York, 1981.
11. **Roberts, J. K. M. and Jardetzky, O.,** Monitoring of cellular metabolism by NMR, *Biochim. Biophys. Acta*, 639, 53, 1981.

12. **Gordon, R. E.,** Magnets, molecules and medicine, *Phys. Med. Biol.,* 30, 741, 1985.
13. **Jacobson, L. and Cohen, J. S.,** Intracellular pH measurements by NMR methods, in *Noninvasive Probes of Tissue Metabolism,* Cohen, J. S., Ed., John Wiley & Sons, New York, 1982, chap. 1.
14. **Williams, S. R. and Gadian, D. G.,** Tissue metabolism studies in vivo by NMR, *Q. J. Exp. Physiol.,* 71, 335, 1986.
15. **Chance, B.,** Is there proton noise in mitochondrial membranes leading to chance fluctuation in pH?, *Nature,* 214, 399, 1967.
16. **Williams, S. R., Gadian, D. G., Proctor, E., Sprague, D. B., Talbot, D. F., Young, I. R., and Brown, F. F.,** Proton NMR studies of muscle metabolites in vivo, *J. Magn. Reson.,* 63, 406, 1985.
17. **Behar, K. L., den Hollander, J. A., Stromski, M. E., Ogino, T., Shulman, R. G., Petroff, O. A. C., and Prichard, J. W.,** High-resolution ^{1}H NMR study of cerebral hypoxia in vivo, *Proc. Natl. Acad. Sci. U.S.A.,* 80, 4945, 1983.
18. **Brown, F. F., Campbell, I. D., Kuchel, P. W., and Rabenstein, D. L.,** Human erythrocyte metabolism studies by ^{1}H spin echo NMR, *FEBS Lett.,* 82, 12, 1977.
19. **Rabenstein, D. L. and Isab, A. A.,** Determination of the intracellular pH of intact erythrocytes by proton NMR spectroscopy, *Anal. Biochem.,* 121, 423, 1982.
20. **Yoshizaki, K., Seo, Y., and Nishikawa, H.,** High-resolution protein magnetic resonance spectra of muscle, *Biochim. Biophys. Acta.,* 678, 283, 1981.
21. **Arus, C., Barany, M., Westler, W. M., and Markley, J. L.,** ^{1}H NMR studies of intact muscle at 11 Tesla, *FEBS Lett.,* 165, 231, 1984.
22. **Chapman, B. E., Kirk, K., and Kuchel, P. W.,** Bicarbonate exchange kinetics at equilibrium across the erythrocyte membrane by ^{13}C NMR, *Biochem. Biophys. Res. Commun.,* 136, 266, 1986.
23. **Balaban, R. S. and Knepper, M. A.,** Nitrogen-14 NMR spectroscopy of mammalian tissues, *Am. J. Physiol.,* 245, C 439, 1983.
24. **Legerton, T. L., Kanamori, K., Weiss, R. L., and Roberts, J. D.,** Measurements of cytoplasmic and vacuolar pH in Neurospora using ^{15}N NMR spectroscopy, *Biochemistry,* 22, 899, 1983.
25. **Taylor, J. S., Deutsch, C., McDonald, G. G., and Wilson, D. F.,** Measurement of transmembrane pH gradients in human erythrocytes using ^{19}F NMR, *Anal. Biochem.,* 114, 415, 1981.
26. **Deutsch, C., Taylor, J. S., and Wilson, D. F.,** Regulation of intracellular pH by human peripheral blood lymphocytes as measured by ^{19}F NMR, *Proc. Natl. Acad. Sci. U.S.A.,* 79, 7944, 1982.
27. **Taylor, J. S. and Deutsch, C.,** Fluorinated alpha-methylamino acids as ^{19}F NMR indicators of intracellular pH, *Biophys. J.,* 43, 261, 1983.
28. **Civan, M. M., Lin, L. E., Peterson-Yantorno, K., Taylor, J., and Deutsch, C.,** Intracellular pH of perfused single frog skin: combined ^{19}F and ^{31}P NMR analysis, *Am. J. Physiol.,* 16, C506, 1984.
29. **Deutsch, C., Taylor, J. S., and Prize, M.,** pH homeostasis in human lymphocytes— modulation by ions and mitogen, *J. Cell Biol.,* 98, 885, 1984.
30. **Kashiwagura, T., Deutsch, C., Taylor, J., Erecińska, M., and Wilson, D. F.,** Dependence of gluconeogenesis, urea synthesis, and energy metabolism of hepatocytes on intracellular pH, *J. Biol. Chem.,* 259, 237, 1984.
31. **Cady, E. B., Delpy, D. T., and Tofts, P. S.,** *Physical Principles and Clinical Applications of NMR,* Lerski, R. A., Ed., Hospital Physicists' Association, London, 1985, 97.
32. **Stewart, I. M., Chapman, B. E., Kirk, K., Kuchel, P. W., Lovric, V. A., and Raftos, J. E.,** Intracellular pH in stored erythrocytes: refinement and further characterization of the ^{31}P NMR methylphosphonate procedure, *Biochim. Biophys. Acta,* 885, 23, 1986.
33. **Labotka, R. J.,** Measurement of intracellular pH and deoxyhemoglobin concentration in deoxygenated erythrocytes by ^{31}P NMR, *Biochemistry,* 23, 5549, 1984.
34. **Labotka, R. J. and Kleps, R. A.,** A phosphate analogue probe of red cell pH using ^{31}P NMR, *Biochemistry,* 22, 6089, 1983.
35. **Navon, G., Ogawa, S., Shulman, R. G., and Yamane, T.,** ^{31}P NMR studies of Ehrlich ascites tumor cells, *Proc. Natl. Acad. Sci. U.S.A.,* 74, 87, 1977.
36. **Civan, M. M., Williams, S. R., Gadian, D. G., and Rozengurt, E.,** ^{31}P NMR analysis of intracellular pH of Swiss Mouse 3T3 cells, *J. Membr. Biol.,* 94, 55, 1986.
37. **Iles, R. A., Griffiths, J. R., Stevens, A. N., Gadian, D. G., and Porteous, R.,** Effects of fructose on the energy metabolism and acid-base status of the perfused-starved rat liver, *Biochem. J.,* 192, 191, 1980.
38. **Enami, I., Akutsu, H., and Kyogoku, T.,** Intracellular pH regulation in acidophilic unicellular alga, *Plant Cell Physiol.,* 27, 1351, 1986.
39. **Seo, Y., Murakami, M., Watari, H., Imai, Y., Yoshizaki, K., Nishikawa, H., and Morimoto, T.,** Intracellular pH determination by a ^{31}P-NMR technique. The second dissociation constant of phosphoric acid in a biological system, *J. Biochem.,* 94, 729, 1983.
40. **Bernard, M., Menasche, P., Canioni, P., Fontanarava, E., Geyer, R. P., Piwnica, A., and Cozzone, P.,** ^{31}P NMR study of high-energy phosphorylated compounds metabolism and intracellular pH in the perfused rat heart, *Arch. Int. Physiol. Biochim.,* 93, 973, 1985.

41. **Adam, W. R., Koretzky, A. P., and Weiner, W.,** ^{31}P NMR in vivo measurement of renal intracellular pH: effects of acidosis and K^+ depletion in rats, *Am. J. Physiol.*, 251, F 904, 1986.
42. **Ackerman, J. J. H., Lowry, M., and Radda, G. K.,** The role of intrarenal pH in regulation of ammoniagenesis: ^{31}P NMR studies of the isolated perfused rat kidney, *J. Physiol.*, 319, 65, 1981.
43. **Burt, C. T., Glonek, T., and Barany, M.,** Analysis of phosphate metabolites, the intracellular pH, and the state of adenosine triphosphate in intact muscle by phosphorus NMR, *J. Biol. Chem.*, 251, 2584, 1976.
44. **Schmidt, H. C., Gooding, C. A., and James, T. L.,** *In vivo* ^{31}P NMR of the liver in the infant rabbit to study the effect of hypoxia on the phosphorus metabolites and pH_{in}, *Invest. Radiol.*, 21, 156, 1986.
45. **Martin, J. B., Bligny, R., Rebeille, F., Douce, R., Leguay, J. J., Mathieu, Y., and Guern, J.,** A ^{31}P NMR study of intracellular pH of plant cells cultivated in liquid medium, *Plant Physiol.*, 70, 1156, 1982.
46. **Moore, R. D. and Gupta, R. K.,** Effect of insulin on intracellular pH as observed by ^{31}P NMR spectroscopy, *Intl. J. Quantum Chem.*, 7, 83, 1980.
47. **Hirose, T., Ueda, T., and Kobatake, Y.,** Changes in intracellular pH accompanying chemoreception in the plasmodia of physarum polycephalum, *J. Gen. Microbiol.*, 128, 2647, 1982.
48. **Jacobson, L. and Cohen, J. S.,** Improved technique for investigation of cell metabolism by ^{31}P NMR spectroscopy, *Biosci. Rep.*, 1, 141, 1981.
49. **Arruda, J. A. L., Wheeler, R. P., Dytko, G., and Talor, Z.,** Intracellular pH of the turtle bladder assessed with fluorescent probes, *Mineral Electrolyte Metab.*, 13, 104, 1987.
50. **Navon, G., Shulman, R. G., Yamane, T., Eccleshall, T. R., Lam, K. B., Baranofsky, J. J., and Marmur, J.,** ^{31}P NMR studies of wild-type and glycolytic pathway mutants of Saccharomyces cerevisiae, *Biochemistry*, 18, 4487, 1979.
51. **Gadian, D. G., Radda, G. K., Dawson, M. J., and Wilkie, D. G.,** pH_i measurements of cardiac and skeletal muscle using ^{31}P-NMR, in *Intracellular pH: Its Measurement, Regulation and Utilization in Cellular Functions,* Nuccitelli, R. and Deamer, D. W., Eds., Alan R. Liss, New York, 1982, 61.
52. **Gillies, R. J., Alger, J. R., den Hollander, J. A., and Shulman, R. G.,** Intracellular pH measured by NMR: methods and results, in *Intracellular pH: Its Measurement, Regulation and Utilization in Cellular Functions,* Nuccitelli, R. and Deamer, D. W., Eds., Alan R. Liss, New York, 1982, 79.
53. **Pollard, H. B., Shindo, H., Creutz, C. E., Pazoles, C. J., and Cohen, J. S.,** *J. Biol. Chem.*, 254, 1170, 1979.
54. **Raftos, J. E., Chapman, B. E., Kuchel, P. W., Lovric, V. A., and Stewart, I. M.,** Intra- and extraerythrocyte pH at 37°C and during long term storage at 4°C: ^{31}P NMR measurements and an electrochemical model of the system, *Haematologia*, 19, 21, 1986.
55. **Slonczewski, J. L., Rosen, B. P., Alger, J. R., and McNab, R. M.,** pH homeostasis in E. coli: measurement by ^{31}P NMR of methylphosphonate and phosphate, *Proc. Natl. Acad. Sci. U.S.A.*, 78, 6271, 1981.
56. **Lin, L. E., Shporer, M., and Civan, M. M.,** ^{31}P NMR analysis of perfused single frog skins, *Am. J. Physiol.*, 248, C 177, 1985.
57. **Bailey, I. A., Williams, S. R., Radda, G. K., and Gadian, D. G.,** Activity of phosphorylase in total global ischaemia in the rat heart: a ^{31}P NMR study, *Biochem. J.*, 196, 171, 1981.
58. **Ugurbil, K., Guernsey, D. L., Brown, T. R., Glynn, P., Tobkes, N., and Edelman, I. S.,** ^{31}P NMR studies of intact anchorage dependent mouse embryo fibroblasts, *Proc. Natl. Acad. Sci. U.S.A.*, 78, 4843, 1981.
59. **Gordon, R. E., Hanley, P. E., and Shaw, D.,** Topical magnetic resonance, *Prog. Nucl. Magn. Reson. Spectrosc.*, 15, 1, 1982.
60. **Gordon, R. E., Hanley, P. E., Shaw, D., Gadian, D. G., Radda, G. K., Styles, P., Bore, P. J., and Chan, L.,** Localization of metabolites in animals using ^{31}P topical magnetic resonance, *Nature*, 287, 736, 1980.
61. **Ackerman, J. J. H., Grove, T. H., Wong, G. G., Gadian, D. G., and Radda, G. K.,** Mapping of metabolites in whole animals by ^{31}P NMR using surface coils, *Nature*, 283, 167, 1980.
62. **Kay, R. R., Gadian, D. G., and Williams, S. R.,** Intracellular pH in Dictyostelium: a ^{31}P NMR study of its regulation and possible role in controlling cell differentiation, *J. Cell Sci.*, 83, 165, 1986.
63. **Chapman, B. E., Kuchel, P. W., Lovric, V. A., Raftos, J. E., and Stewart, I. M.,** Regeneration of phosphorylated metabolites in stored erythrocytes in an open perfusion system: studies using ^{31}P NMR spectroscopy, *Br. J. Haematol.*, 61, 385, 1985.
64. **Desmoulin, F., Cozzone, P. J., and Cancioni, P.,** ^{31}P NMR study of phosphorylated metabolites compartmentation, intracellular pH and phosphorylation in the perfused rat liver, *Eur. J. Biochem.*, 162, 151, 1987.
65. **Arruda, J. A. L., Lubansky, H., Dytko, G., Mola, R., Kleps, R. A., and Burt, C. T.,** Intracellular pH of the urinary bladder and red blood cells of the fresh water turtle, *Mineral Electrolyte Metabol.*, 10, 233, 1984.

66. **Arruda, J. A. L., Lubansky, H., Sabatini, S., Dytko, G., and Burt, C. T.,** Intracellular phosphate pool show compartmentalization of intracellular pH in the turtle bladder, *Biochem. Biophys. Res. Commun.,* 100, 1459, 1981.
67. **Roberts, J. K. M., Wade-Jardetzky, N., and Jardetzky, O.,** Intracellular pH measurement by ^{31}P NMR. Influence of factors other than pH on ^{31}P chemical shifts, *Biochemistry,* 20, 5389, 1981.
68. **Hope, P. L., Cady, E. B., Chu, A., Delpy, D. T., Gardiner, R. M., and Reynolds, E. O. R.,** Brain metabolism and intracellular pH during ischaemia and hypoxia: an in vivo ^{31}P and ^{1}H NMR study in the lamb, *J. Neurochem.,* 49, 75, 1987.
69. **Satre, M., Klein, G., and Martin, J. B.,** Intracellular pH control in Dictyostelium discoideum: a ^{31}P-NMR analysis, *Biochimie,* 68, 1253, 1986.
70. **Satre, M. and Martin, J. B.,** ^{31}P-NMR analysis of the intracellular pH in the slime mold Dictyostelium discoideum, *Biochem. Biophys. Res. Commun.,* 132, 140, 1985.
71. **Morrill, G. A., Kostellow, A. B., Mahajan, S., and Gupta, R. K.,** Role of calcium in regulating intracellular pH following the stepwise release of the metabolic blocks at first-meiotic prophase and second-meiotic metaphase in amphibian oocytes, *Biochim. Biophys. Acta,* 804, 107, 1984.
72. **Morrill, G. A., Weinstein, S. P., Kostellow, A. B., and Gupta, R. K.,** Studies of insulin action on the amphibian oocyte plasma membrane using NMR, electrophysiological and ion flux techniques, *Biochim. Biophys. Acta,* 844, 377, 1985.
73. **Griffiths, J. R. and Iles, R. A.,** NMR studies of tumours, *Biosci. Rep.,* 2, 719, 1982.
74. **Roberts, J. K. M., Callis, J., Wemmer, D., Walbot, V., and Jardetzky, O.,** Mechanism of cytoplasmic pH regulation in hypoxic maize root tips and its role in survival under hypoxia, *Proc. Natl. Acad. Sci. U.S.A.,* 81, 3379, 1984.
75. **Lewis, S. F., Haller, R. G., Cook, J. D., and Nunnally, R. L.,** Muscle fatigue in McArdle's disease studied by ^{31}P-NMR: effect of glucose infusion, *J. Appl. Physiol.,* 59, 1991, 1985.
76. **Roberts, J. K. M., Wemmer, D., Ray, P. M., and Jardetzky, O.,** Regulation of cytoplasmic and vacuolar pH in maize root tips under different experimental conditions, *Plant Physiol.,* 69, 1344, 1982.
77. **Kime, M. J., Ratcliffe, R. G., Williams, R. J. P., and Loughman, B. C.,** The application of ^{31}P NMR to higher plant tissue. I. Detection of spectra, *J. Exp. Bot.,* 33, 656, 1982.
78. **Kime, M. J., Ratcliffe, R. G., and Loughman, B. C.,** The application of ^{31}P NMR to higher plant tissue. II. Detection of intracellular changes, *J. Exp. Bot.,* 33, 670, 1982.
79. **Jacobus, W. E., Kallman, C. H., Weisfeldt, M. L., and Flaherty, J. T.,** NMR studies of intracellular pH and myocardial contractility during ischemia, in *Cardiology,* Chazov, E. I., Smirnov, V. N., and Oganov, R. G., Eds., Plenum Press, New York, 1984.
80. **Ballarin-Denti, A., den Hollander, J. A., Sanders, D., Slayman, C. W., and Slayman, C. L.,** Kinetics and pH-dependence of glycine-proton symport in Saccharomyces cerevisiae, *Biochim. Biophys. Acta,* 778, 1, 1984.
81. **Winkler, M. M., Matson, G. B., Hershey, J. W. B., and Bradbury, E. M.,** ^{31}P-NMR study of the activation of the sea urchin egg, *Exp. Cell Res.,* 139, 217, 1982.
82. **Koeze, T. H., Lantos, P. L., Iles, R. A., and Gordon, R. E.,** *In vivo* NMR spectroscopy of a transplanted brain tumor, *Br. J. Cancer,* 49, 357, 1984.
83. **Dawson, M. J. and Wray, S.,** The effects of pregnancy and parturition on phosphorus metabolites in rat uterus studied by ^{31}P NMR, *J. Physiol.,* 368, 19, 1985.
84. **Nunnally, R. L., Stoddard, J. S., Helman, S. I., and Kokko, J. P.,** Response of ^{31}P-NMR spectra of frog skin to variations in P_{CO_2} and hypoxia, *Am. J. Physiol.,* 245, F 792, 1983.
85. **Tanokura, M. and Yamada, K.,** Changes in intracellular pH and inorganic phosphate concentration during and after muscle contraction as studied by time-resolved ^{31}P-NMR, *FEBS Lett.,* 171, 165, 1984.
86. **Hamm, J. R. and Yue, G. M.,** ^{31}P NMR measurements of intracellular pH in giant barnacle muscle, *Am. J. Physiol.,* 252, C30, 1987.
87. **Ruigrok, T. J. C., Kirkels, J. H., van Echteld, C. J. A., Borst, C., and Meijler, F. L.,** ^{31}P NMR study of intracellular pH during the calcium paradox, *J. Mol. Cell. Cardiol.,* 19, 135, 1987.
88. **Jans, A. W. H., Amsler, K., Griewel, B., and Kinne, R. K. H.,** Regulation of intracellular pH in LLC-PK_1 cells studied using ^{31}P NMR spectroscopy, *Biochim. Biophys. Acta,* 927, 203, 1987.
89. **Nioka, S., Chance, B., Hilberman, M., Subramanian, H. V., Leigh, J. S., Jr., Veech, R. L., and Forster, R. E.,** Relationship between intracellular pH and energy metabolism in dog brain measured by ^{31}P NMR, *J. Appl. Physiol.,* 62, 2094, 1987.
90. **Benn, R. and Günter, H.,** Modern pulse methods in high-resolution NMR spectroscopy, *Angew. Chem. Int. Ed. Engl.,* 22, 350, 1983.
91. **Ogawa, S., Shulman, R. G., Glynn, P., Yamane, T., and Navon, G.,** On the measurement of pH in *Escherichia coli* by ^{31}P NMR, *Biochim. Biophys. Acta,* 502, 45, 1978.
92. **Fossel, E. T., Morgan, H. E., and Ingwall, J. S.,** Measurement of changes in high energy phosphates in the cardiac cycle by using gated ^{31}P NMR, *Proc. Natl. Acad. Sci. U.S.A.,* 77, 3654, 1980.

93. **Wemmer, D., Wade-Jardetzky, N., Robin, E., and Jardetzky, O.,** Changes in the phosphorus metabolism of a diving turtle observed by ^{31}P NMR, *Biochim. Biophys. Acta,* 720, 281, 1982.
94. U.S. Food and Drug Administration, *Guidelines for Evaluating Electromagnetic Exposure Risks for Trials of Clinical NMR Systems,* Rockville, MD, 1982.
95. NRPB, Exposure to nuclear magnetic resonance clinical imaging radiography, 47, 258, 1981.
96. **Gullino, P. M. and Knazek, R. A.,** Tissue culture on artificial capillaries, *Methods Enzymol.,* 58, 178, 1979.
97. **Cohen, J. S.,** ^{31}P NMR studies of cell metabolism, in *Magnetic Resonance in Biology and Medicine,* Govil, Khatrapal, and Saran, Eds., McGraw-Hill, New Delhi, 1985, 77.
98. **Lyon, R. C., Faustino, P. J., and Cohen, J. S.,** A perfusion technique for ^{13}C NMR studies of the metabolism of ^{13}C-labelled stubstates by mammalian cells, *Magn. Res. Med.,* 4, 24, 1987.
99. **Gonzales-Mendez, R., Wemmer, D., Hahn, G., Wade-Jardetzky, N., and Jardetzky, O.,** Continuous-flow NMR culture system for mammalian cells, *Biochim. Biophys. Acta,* 720, 274, 1982.
100. **Mattiasson, B., Ed.,** *Immobilized Cells and Organelles,* Vols. I and II, CRC Press, Boca Raton, FL, 1983.
101. **Katz, R., Yeh, H. J. C., and Johnson, D. F.,** A quantitative study of organic phosphate-amine interaction by ^{31}P NMR, *Biochem. Biophys. Res. Commun.,* 58, 316, 1974.

Chapter 5

THE CHOICE OF THE ASSAY METHOD

Jan Slavík

TABLE OF CONTENTS

I. INTRODUCTION

Comparison of different intracellular pH-measuring techniques is rather complicated due to the fact that they are based on quite different physical principles. The technical advances of the past decade, together with a steady search for more reliable pH_{in} measuring techniques, have led to the introduction of several new techniques, such as NMR spectroscopy and fluorescence probe pH topography. This development is documented in a number of reviews on intracellular pH measurement.[1-14]

II. THE "AVERAGE" INTRACELLULAR pH AS SEEN BY DIFFERENT TECHNIQUES

Due to the heterogeneity of intracellular pH and the individual differences between cells in a population, each particular technique yields a different kind of average intracelular pH value:

- The distribution techniques, for instance, yield lower pH readings if a weak base is used than if a weak acid is employed, thus, bracketing the volume-averaged pH value (cf. Chapter 4B. I. C.).
- The pH value of a cell-free homogenate reflects the relative buffering powers of various intracellular compartments together with their pH.
- The spectroscopic techniques yield a weighted average with respect to the intracellular distribution on the indicator and the shape of the pH standard curve. In this context, the fluorescence-probe pH topography is the only technique that produces two-dimensional maps with pH values averaged only in the vertical direction.
- The microelectrodes apparently provide the lcoal pH value of the spot they are inserted into (with respect to the size of the tip and its insulation).

III. COMPARISON OF TECHNIQUES: THEIR ADVANTAGES AND DISADVANTAGES

None of the presently used intracellular pH measurements is universally applicable, and none gives entirely reliable pH values. The techniques differ in the degree of interference (invasiveness) as well as in their complexity. Their advantages and drawbacks may be summarized as follows.

A. Distribution Techniques

Advantages of the distribution techniques are the simplicity of their execution and their reliability. The disadvantages are, especially, the slow response (which is related to the volume-to-surface ratio, thus, slower with larger cells). The time required for equilibration ranges from seconds for organelles to minutes or tens of minutes for some especially recalcitrant large cells. Quite often, only a single measurement per sample is possible, and the measured intracellular pH may be influenced by the very measurement procedure (the presence of weak acid or base).

B. Microelectrodes

Microelectrodes require larger cells which withstand the impalement of the electrode(s). Their advantage lies in a simple, straightforward calibration, long-term stability, and the possibility of continuous reading of intracellular values. In all cases, only one individual cell is investigated at a time. The drawback of this technique is the uncertainty whether or not the microelectrode will form a perfect seal with the membrane. This is not always achieved, and leaks of ions around the microelectrode tip will cause the measured potential to drop in an ill-controlled way.

C. Absorption

Absorption techniques allow either cell populations or individual cells to be measured. They are simple and permit continuous detection as well as detection of rapid pH changes. Their disadvantage may lie in a rather complicated calibration, and, in some cases, a leak of the dye from cells occurs. Also, an intense coloring of cells (e.g., due to hemoglobin) may cause difficulties.

D. Fluorescence

Fluorescence techniques are slightly more complicated than the absorption ones. They offer the same advantages, but, due to their higher sensitivity, they are fit for special procedures, such as pH topography or flow cytometry. Like absorption, they are convenient for monitoring fast pH changes as well as for continuous recording of pH. The drawbacks are more or less the same as those of absorption. The flow cytometry offers the possibility of measurement of individual pH values of a large number of cells (e.g., 1000 to 10,000/s) and may be employed in a cell-sorting device that would separate cells according to their pH values. Intracellular pH topography makes it possible to map local pH values inside the cell and intracellular compartments and, thus, offers a unique possibility to examine intracellular and membrane processes in which the release or consumption of protons are involved.

E. Nuclear Magnetic Resonance

NMR spectroscopy has now become a routine technique for intracellular measurement, though its complexity still limits its applications. In compensation for this, it offers the assessment of the physiological state of cells simultaneously with the pH measurement. The drawback of NMR is the low sensitivity requiring high concentrations of cells (close packing).

IV. COMPARISON OF pH-MEASURING TECHNIQUES WITH RESPECT TO THE PROBLEM STUDIED

A. A Whole Animal, Organ, or a Large Tissue Section

For such cases only NMR yields meaningful average pH values of the whole sample.

B. Cell Populations

When working with cell suspensions or monolayers of cells, the distribution of weak electrolytes gives the average pH value of all cells and compartments. The NMR technique gives much more information in the form of broadening, or even splitting, of peaks. This reflects both the heterogeneity of intracellular pH and the heterogeneity of pH values of individual cells. Optical techniques are also convenient, yielding average intracellular pH values of the cell population.

The existence of special subpopulations of cells in a suspension is most easily followed by flow cytometry.

When working with a tissue, fluorescence or absorption is the method of choice as microspectrophotometric techniques can distinguish between individual cells or neighboring cells.

C. Individual Cells

For the measurement of intracellular pH values of individual cells, only microelectrodes and optical techniques come into consideration. In these experiments, the size of the cell is the most important limiting factor. With microelectrodes, the diameter of the cell should be approximately 10 μm. With fluorescence or absorption it is basically the visibility of the cell under the microscope (some 0.2 μm) which is limiting, though microspectrofluorometry can produce a reading from a cell that cannot be seen by the observer using the microscope.

D. Subcellular pH Measurements

Both microelectrodes and optical techniques can be employed for measurements inside the cell and across intracellular compartments. The spatial resolution of microelectrodes is limited by the size of their tip (1 μm); the spatial resolution of optical techniques, by the wavelength of the light used (i.e., about 0.15 to 0.20 μm in the optical microscope). The other pH-measuring techniques cannot be used, with the exception of suspensions of separated organelles which can be treated as suspensions of cells.

E. Temporal Resolution — Rapid Changes

The highest temporal resolution can be accomplished by optical techniques where the rate-limiting process is the response of the dye to the ambient pH changes, typically around 30 μs.

The distribution techniques allow a time resolution in seconds if organelle pH is measured and tenths of seconds, at best, if whole cells are examined. The process of measurement may be speeded up if the rates of influx rather than the equilibrium concentrations are followed.

The time response of pH-sensitive microelectrodes is typically 5 to 30 s.

The temporal resolution of the NMR techniques is generally in the range of minutes because of the low sensitivity of NMR, but, in the case of cyclic events, it may be increased to seconds.

F. Temporal Resolution — Continuous Monitoring over a Longer Time Period

At the other end of the time resolution spectrum, it is the use of microelectrodes and optical techniques which is the most convenient and generally applicable. The optical techniques may sometimes be invalidated by bleaching or leakage of the dye. Nevertheless, they represent the best choice for continual monitoring of cell cultures. The NMR techniques are suitable for whole animals or organs but not for cell cultures, as the survival of densely packed cells may be difficult.

The distribution techniques can be adapted to continuous measurement by a continuous recording of external concentration of the acid or base using an ion-selective electrode, flow dialysis, or by optical monitoring of extracellular concentrations of dyes.

The use of microelectrodes is also quite convenient for long-term observation as electrodes usually exhibit a fair long-term stability (certainly higher than the impaled cells would).

G. Sensitivity

Optical techniques and microelectrodes require only one cell for measurement. The distribution techniques require some 100 nl of cells (i.e., about 10^7 cells).

^{31}P NMR needs about 10^4 times more cells (as corresponds to 1 μmol P_i in approximately 10 ml of a 20 to 50% suspension).

H. Sensitivity to Minute pH Changes

The sensitivity for recording small, transient pH changes is probably highest in the case of absorption and fluorescence techniques, where pH changes as low as 0.01 units can normally be followed without special effort. This sensitivity, together with the high temporal resolution, makes optical techniques ideal for monitoring intracellualr pH changes.

REFERENCES

1. **Caldwell, P. C.,** Intracellular pH, *Int. Rev. Cytol.,* 5, 229, 1956.
2. **Robson, J. S., Bone, J. M., and Lambie, A. T.,** Intracellular pH, in *Advances in Clinical Chemistry,* Vol. 11, Bodansky, O. and Stewart, C. P., Eds., Academic Press, New York, 1968, 213.
3. **Waddell, W. J. and Bates, R. G.,** Intracellular pH, *Physiol. Rev.,* 49, 285, 1969.
4. **Cohen, R. D. and Iles, R. A.,** Intracellular pH, measurement, control and metabolic interrelationships, *Crit. Rev. Clin. Lab. Sci.,* 6, 101, 1975.
5. **Boron, W. F. and Roos, A.,** Comparison of microelectrode, DMO and methylamine methods for measuring intracellular pH, *Am. J. Physiol.,* 231, 799, 1976.
6. **Poole-Wilson, P. A.,** Measurement of myocardial intracellular pH in pathological states, *J. Mol. Cell. Cardiol.,* 10, 512, 1978.
7. **Gillies, R. J. and Deamer, D. W.,** Intracellular pH, methods and applications, in *Cell Cycle Regulation,* Academic Press, New York, 1978, 105.
8. **Rottenberg, H.,** The measurement of membrane potential and pH gradient in cells, organelles and vesicles, *Methods Enzymol.,* 55, 547, 1979.
9. **Radda, G. K. and Seeley, P. J.,** *Annu. Rev. Physiol.,* 41, 749, 1979.
10. **Roos, A. and Boron, W. F.,** Intracellular pH, *Physiol. Rev.,* 61, 296, 1981.
11. **Iles, R.,** Measurement of intracellular pH, *Biosci. Rep.,* 1, 687, 1981.
12. **Marx, J. L.,** Investigators focus on intracellular pH, *Science,* 213, 745, 1981.
13. **Nuccitelli, R. and Deamer, D. W., Eds.,** *Intracellular pH: Its Measurement, Regulation and Utilization in Cellular Functions,* Alan R. Liss, New York, 1982.
14. **Gillies, R. J. and Deamer, D. W.,** Intracellular pH: methods and applications, *Curr. Top. Bioenerg.,* 9, 63, 1979.

Chapter 6

HETEROGENEITY OF INTRACELLULAR pH

Arnošt Kotyk and Jan Slavík

TABLE OF CONTENTS

I. INTRODUCTION

The idea that intracellular pH represents only an approximation, frequently a crude one, winds its way through this book from beginning to end. The living cell exhibits heterogeneity in intracellular pH due both to internal compartmentation and to pH differences within the compartments.

Attempts to define intracellular regions of different pH have been many, and it has, indeed, been possible to assign a distinct pH to plant or yeast vacuoles and even to mitochondria. The observation that digestive intracellular organelles may be more acid than the cytosol is now considered an established fact (Figure 79), and hydrogen ion pumps are known to operate in their membranes, with the obvious purpose of keeping the intraorganellar pH low, which is helpful to their hydrolytic function.

Nonuniformity of pH, even within defined morphological compartments is another matter, and its fully satisfactory explanation is still lacking. Let us first summarize the evidence that has accumulated over the years in this respect.

A. Data from Microelectrode Measurements

Discussions on proton distribution across the membrane were initiated by Carter et al. in 1967.[1] Their experiments were with two microelectrodes, the larger Caldwell type supposedly penetrating deeper into the cell (barnacle muscle fibers about 1 mm in diameter). This gave a reading of pH 7.0, while the smaller, glaze-insulated electrode gave only a pH of 6.1 if inserted just under the cell surface. A pH measurement based on distribution of a labeled compound (hence, an ''average'' value) resulted in a pH of 6.7. On the basis of these results, it was postulated[2] that there is a thin, acidic layer comprising about 3% of the total cell volume, located immediately beneath the membrane.

These results were later confirmed[3] in frog sartorius muscle. Similarly, values of pH 6.1 were registered just as the electrodes were introduced into a cell, but the pH was 6.7, as the electrodes were farther advanced into the rat or crab muscle fiber.

Still, a later discussion of these experiments[4,5] favored the view that an inadequate insulation of the electrode tip was the underlying cause, rather than the actual pH difference.

Recent measurements of extracellular pH values with single- and double-barreled glass microelectrodes[6,8] ahowed significant differences between the bulk extracellular value (pH_{out}), the value close to the external surface ($pH_{surface}$), and the intracellular value (pH_{in}) of various muscle cells. The difference between pH_{out} and $pH_{surface}$ was more pronounced in media with a lower buffering power (20 m*M* HEPES yielded a difference of 0.15 pH; 2 m*M* HEPES, 0.6 to 0.7 pH units). The $pH_{surface}$ values always lay between pH_{out} and pH_{in}. A constant bulk pH was only found some 300 μm from the membrane.

B. NMR Estimations

The NMR peaks of pH-sensitive compounds, particularly the P_i peaks in ^{31}P NMR, are often much broader than those of the pH-insensitive ones. The broadening may arise from the variation of pH between cells, between cell compartments, and within compartments. Usually, if two or three clearly resolvable lines are obtained in the spectrum, they are attributed to a pH-indicating compound at two distinct pH values of two different compartments (mostly a cytoplasmic and a vacuolar peak). The appearance of a broad resonance is interpreted as several overlapping lines due to P_i experiencing a broad range of different pH values.

Roberts and Jardetzky[9] compared data on line-width and T_2 values of orthophosphate and phosphocreatine obtained at different intensities of the magnetic field. They concluded that the results are consistent with the view that the chemical shift heterogeneity and the consequent broad resonance are probably caused by pH nonuniformity within the cell.

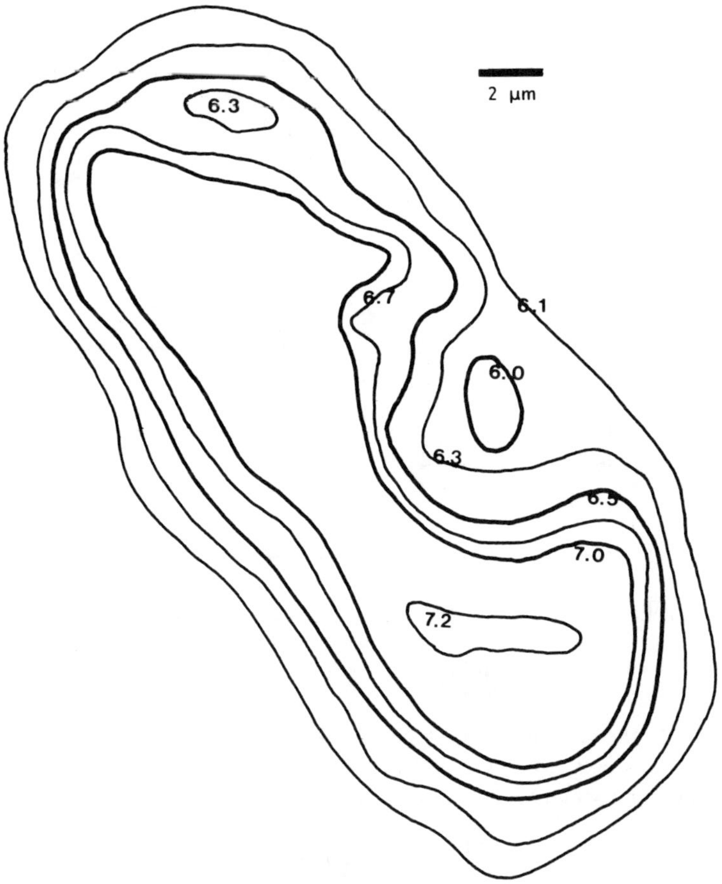

FIGURE 79. pH contour lines of a cell of *Endomyces magnusii* determined from the fluorescence of intracellular fluorescein following excitation with two different wavelengths. The region at center right corresponds to an accumulation of several vacuoles.

C. Absorption Assays

Visual microscopic observation of stained cells showed different colors at different regions inside the cell. For instance, injection of bromothymol blue into muscle fibers[2] seems to confirm the microelectrode findings that deeper layers are more alkaline than those close to the membrane.

Using neutral red, Schmidtmann[10,11] observed a thin layer of cytoplasm surrounding the nucleus which was more acidic than the rest of the cytoplasm and which disappeared when the nuclear membrane was destroyed.

D. Fluorescence Observations

Microspectrofluorometry of fluorescein-filled amebas (with a spot diameter of some 45 μm) showed the same pH reading for repeated measurement of the same area, but two different areas within the same cell differed by as much as 0.4 pH units, suggesting a significant intracytoplasmic pH heterogeneity.[12]

The fluorescence-based technique of intracellular pH topography showed differences in pH both between different compartments and within a compartment (cf. Figure 79). This was found both in yeasts[13] and in the fungus *Penicillium cyclopium*.[14]

From a scrutiny of all these results, it would appear that the local pH value deviates most

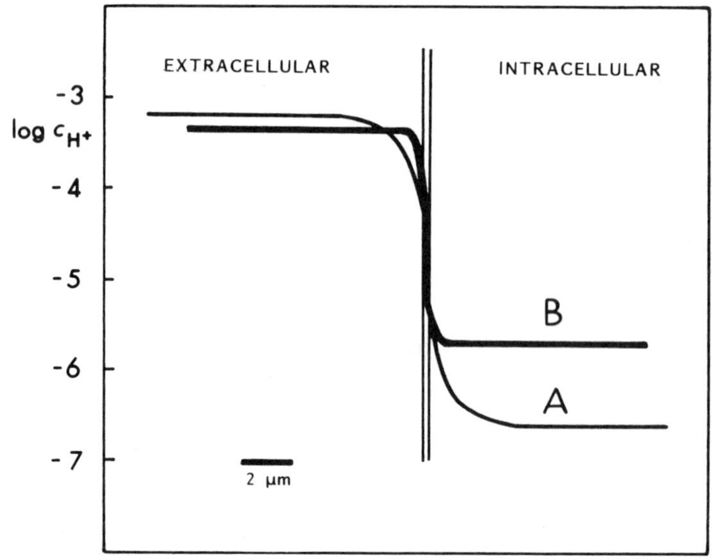

FIGURE 80. pH profiles across the plasma membrane of a yeast cell which H^+ ions move passively. Curve A corresponds to suspension in a dilute acid; curve B, in a strong buffer of pH 3.3.

at, or near, cell or organelle membranes, across which H^+ ions are transported. Most detailed data on this phenomenon are available from yeast,[13,15] and these will be used to discuss the relevance of this microheterogeneity of pH values.

II. CONCLUSIONS

First, the H^+ ions simply permeate without the activity of a pump (or through a pump, downhill), such as is the case, for instance, when microbial cells are suspended in an acid (pH 3) medium. Here, one would expect a smooth concentration profile across the membrane (Figure 80) caused by the fact that diffusion of H^+ ions toward and away from the membrane — although rapid ($D_{H^+} = 9.3 \cdot 10^{-9}$ m² s^{-1}, as compared with self-diffusion of water with $D_{H_2O} = 2.3 \cdot 10^{-9}$ m² s^{-1}, or diffusion or other ions, with D_{Na^+} $1.3 \cdot 10^{-9}$ m²s^{-1} and D_{Cl^-} $2.0 \cdot 10^{-9}$ m² s^{-1}) — still does not measure up to the transmembrane leak of H^+. Therefore, very small stagnant or unstirred layers are formed even with respect to H^+.[15] The effective thickness of these unstirred layers is greater the lower the diffusion coefficient of the permeating substance and the higher the rate of its transmembrane movement.

If the cells are put in a buffer, the gradual character of the transmembrane pH profile soon disappears, and the pH of the entire cell tends to be uniform, even if distinct from the uniform extracellular solution (cf. Figure 80).

Second, the H^+ ions are pumped actively across the membrane in the outward direction. Using similar considerations as above, a somewhat more complicated profile would be expected (Figure 81A). However, even here the measurable profle was smooth as shown by curve B in Figure 81. Moreover, intracellularly the gradual change of pH continued for some 3 to 4 μm so that in a smaller cell, such as that of *Saccharomyces cerevisiae,* pH changed from a value of about 5.5 at the membrane of a cell suspended in water to about 7.0 in its very center. With larger cells, such as *Endomyces magnusii*, there was a relatively alkaline central region some 4 × 10 μm² in area (Figure 82).

Remarkably, this nonhomogeneity is absent in cells that are starved or have inhibited respiration. In such cells, the pH is homogeneous although still distinct from the external pH.

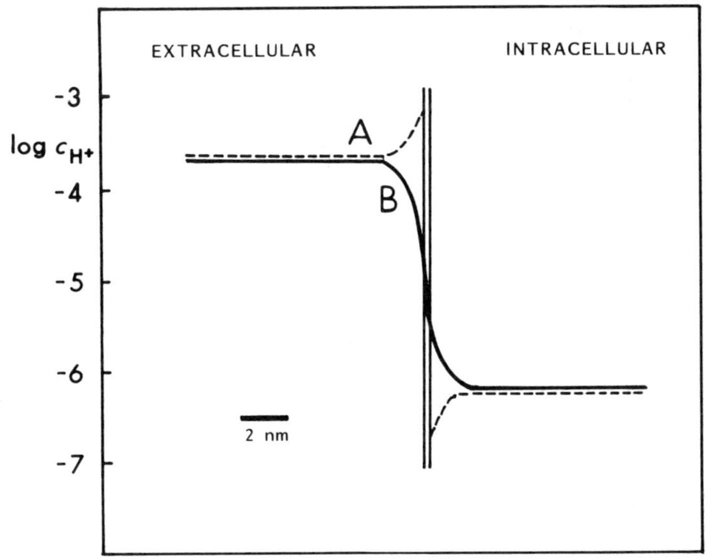

FIGURE 81. Theoretical pH profile across a membrane actively pumping H^+ ions outward (curve A), and the actually found smooth pH profile (curve B).

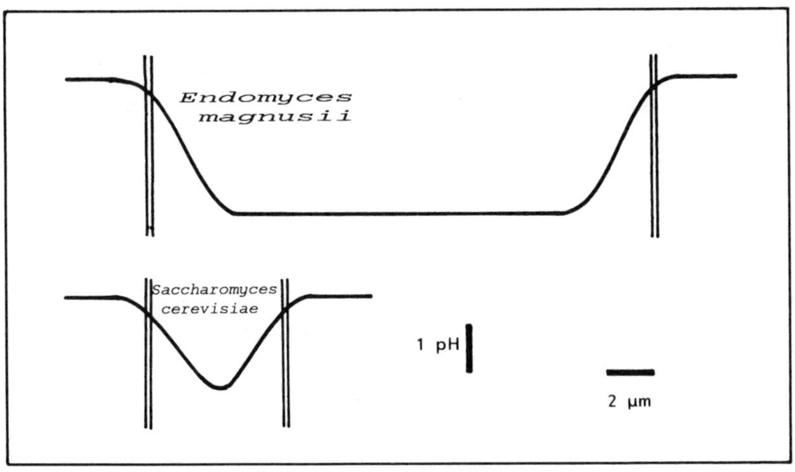

FIGURE 82. Observed pH profiles across the cells of two yeast species suspended in water.

The stability of this centrifugal pH decrease in active cells is particularly apparent if cells are suspended in an alkaline buffer (Figure 83) where the actual pH differences across the membrane may be much greater than would appear from the bulk values.

It is clear that several processes of H^+ ion movement are involved in the peculiarities of perimembrane pH:

1. Formation of stagnant, unstirred layers where H^+ ions move down their concentration gradient across the membrane
2. Action of several H^+ ion "pumps", among them the H^+-translocating plasma membrane ATPase, rapid downhill diffusion of intracellularly generated CO_2, and an H^+/K^+ exchange that can operate in both directions

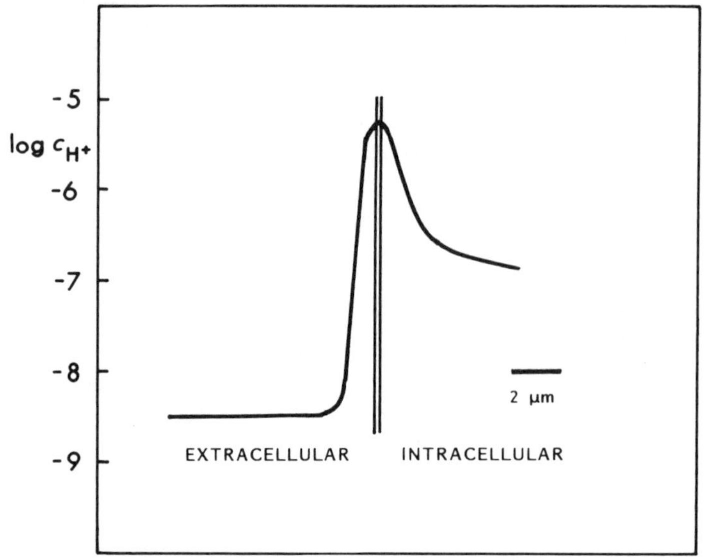

FIGURE 83. The observed pH profile across the plasma membrane of a baker's yeast cell suspended in a dilute buffer of pH 8.5.

3. Role of surface potential which, in response to a generally negative surface charge of membranes, attracts positive ions from the surroundings. (Here, depending on the presence of other cations, the H^+ ions are attracted to the membrane to a greater or lesser degree. This process, however, is limited to a very thin perimembrane layer of only few nanometers thick) three orders of magnitude thinner than the observed area of pH_{in} heterogeneity.

Thus, in spite of the substantial progress made in the last few years in the measurement not only of the bulk, but also of the local pH within cells, a technique with a higher resolution than that imposed by visible light properties will be requried to permit one to speak of pH at a more ''molecular'' level. A quite promising application appears to be that of the use of fluorescein-labeled monoclonal antibodies.

Will a membrane-permeant substance that would form an electron-dense precipitate only below or above a certain pH be available?

REFERENCES

1. **Carter, N. W., Rector, F. C., Campion, R. T., and Seldin, D. W.,** Measurement of intracellular pH of skeletal muscle with pH-sensitive glass microelectrodes, *J. Clin. Invest.*, 46, 920, 1967.
2. **Carter, N. W.,** Intracellular pH, *Kidney Int.*, 1, 341, 1972.
3. **Hannan, S. F. and Wiggins, P. M.,** Intracellular pH of frog sartorius muscle, *Biochim. Biophys. Acta,* 428, 205, 1976.
4. **Roos, A. and Boron, W. F.,** Intracellular pH, *Physiol. Rev.*, 61, 297, 1981.
5. **Cohen, R. D. and Iles, R. A.,** Intracellular pH: measurement, control, and metabolic interrelationship, *Crit. Rev. Clin. Lab. Sci.*, 6, 101, 1975.
6. **Vanheel, B., de Hemptinne, A., and Leusen, I.,** Influence of surface pH on intracellular pH regulation in cardiac and skeletal muscle, *Am. J. Physiol.*, 250, C748, 1986.
7. **Vanheel, B., de Hemptinne, A., and Leusen, I.,** Intracellular pH and contraction of isolated rabbit and cat papillary muscle: Effect of superfusate buffering, *J. Mol. Cell. Cardiol.*, 17, 23, 1985.

8. **de Hemptinne, A. and Huguenin, F.,** The influence of muscle respiration and glycolysis on surface and intracellualr pH in fibres of the rat soleus, *J. Physiol.,* 347, 581, 1984.
9. **Roberts, J. K. M. and Jardetzky, O.,** Monitoring of cellular metabolism by NMR, *Biochim. Biophys. Acta,* 639, 53, 1981.
10. **Schmidtmann, M.,** Über eine Methode zur Bestimmung der Wasserstoffzahl in Gewebe und in einzelnen Zellen, *Biochem. Z.,* 150, 253, 1924.
11. **Schmidtmann, M.,** Über die intracelluläre Wasserstoffionenkonzentration unter physiologischen und einigen pathologischen Bedingungen, *Z. Gesamte Exp. Med.,* 45, 714, 1925.
12. **Heiple, J. M. and Taylor, D. L.,** Intracellular pH in single motile cells, *J. Cell Biol.,* 86, 885, 1980.
13. **Slavík, J. and Kotyk, A.,** Intracellular pH distribution and transmembrane pH profile of yeast cells, *Biochim. Biophys. Acta,* 766, 679, 1984.
14. **Roos, W. and Slavík, J.,** Intracellular pH topography of Penicillium cyclopium protoplasts. Maintenance of ΔpH by both passive and active mechanisms, *Biochim. Biophys. Acta,* 899, 67, 1987.
15. **Slavík, J.,** Intracellular pH topography: determination by a fluorescent probe, *FEBS Lett.,* 156, 227, 1983.

INDEX

A

B

C

D

E

F

G

H

I

O

P

Q